SECRETS OF THE ICE AGE

Also by Evan Hadingham

The Fighting Triplanes
Ancient Carvings in Britain
Circles and Standing Stones

Entrance to Les Pradières, central Pyrenees, France, a cave occupied by
hunters at intervals throughout the last Ice Age.

SECRETS
OF THE
ICE AGE

A Reappraisal of Prehistoric Man

EVAN HADINGHAM

WALKER AND COMPANY
New York

This book is for
Alain, Alex and Antony
with my thanks

See page 332 for photo credits.

First published in the United States of America
in 1979 by the Walker Publishing Company, Inc.

Published simultaneously in Canada by
John Wiley & Sons Canada,
Limited, Rexdale, Ontario.

ISBN: 0-8027-0624-X Cloth
0-8027-7192-0 Paper

Library of Congress Catalog Card Number: 78-65614

Printed in the United States of America

10 9 8 7 6 5 4 3 2

Contents

Foreword

This book began in the summer of 1975 with my experiences as a digger at prehistoric excavations in southwest France. The work involved removing deposits of soil and limestone with extreme care and patience, often in temperatures of over 90°. In these conditions it was extraordinary to come across the bones and antlers of reindeer, which must have existed in an environment utterly different from that of the present day. It was also remarkable to recover the flint tools that had been made at this remote period of prehistory, for the refined shapes and delicately trimmed edges of these implements made it obvious that they were the product of highly skilled craftsmen.

My curiosity about this vanished world grew as I visited the decorated caves of southern France and northern Spain. Like many others before me, I became convinced that the most popular theories of the cave artists' motives were inadequate for they always failed to explain some aspect—either the variety, quality or character—of the prehistoric artworks. The paintings and engravings in each cave were a little different from the last, and yet an underlying unity of style and intention could be sensed wherever one looked.

These were subjective impressions, and, in fact, as later reading suggested, reasonable ideas *do* exist that throw light on the behavior and beliefs of the Ice Age hunters of western Europe. However, much of the evidence lies scattered in the pages of archaeological textbooks or of journals in Spanish or French. An attempt to introduce some of this material to the general reader seemed worthwhile. The recent resurgence of interest in modern hunter-gatherers also seemed too important to ignore, although it is by no means certain that observations made among recent peoples such as the !Kung Bushmen or the Netsilik Eskimo are relevant to the life led by communities ten thousand or more years ago. Could some overall picture of the hunting life, past and present, be drawn from this diverse material?

In any popular writing there is a risk of suggesting too much completeness and accuracy in the state of our knowledge of the past. Many of the ideas explored here, such as the prehistoric domestication of reindeer or the existence of underground "sanctuaries," remain possibilities rather than certainties. It is important for the reader to bear in mind that much of the discussion relates to just one narrow region of Europe. All the dates cited in the text are extremely tentative and should never be read with the confidence of those appearing in a history book. Finally, it must be emphasized that the use of the term "Ice Age" in capital letters throughout this book is a convenient shorthand used to refer only to the period of the last (Würm) Ice Age; this is explained fully in Chapter 2. It is this one glacial period, dating very roughly from 70,000 to 9,000 B.C., that forms the subject of the book.

A full list of acknowledgments can be found at the back of the book, but I must record here my particular debt to Alex Hooper, whose constant encouragement and expert advice were invaluable at every stage. I also owe a debt to the archaeologists M. and Me. F. Champagne and R. Arambourou who, during my periods of work as a volunteer on their excavation teams, first encouraged my interest in the subject.

It would be impossible to thank all the many individuals who made it possible for me to visit decorated caves in southern France, northern Spain and southern Italy, but I particularly appreciate the time and attention given to me by Dr. Jesus Altuna, Jean Clottes, Alain Fournier and Luc Wahl. My thanks also go to my companions on numerous guided explorations underground, Alain and Marie-Claude Fournier, Alex Hooper, Paula Hutchings, Patricia Roberts, Antony Stevens and Clive Williams.

Much of the research for this book was undertaken while the author was a graduate student in the Department of Prehistory and Archaeology, University of Sheffield, under Professor Keith Branigan. The influence of the Department and of its lecturers, especially Dr. Robin Dennell, Andrew Fleming, Dr. Paul Mellars and Robin Torrence, contributed enormously to this book although the responsibility for the speculations offered here is entirely my own. I was able to visit the decorated cave of Levanzo during the Sheffield University survey of central Sicily, led by Dr. Richard Hodges. Thanks also to fellow students Neil Carver, Dave Fine, Tom Levy, Mark Matthewman, Nigel Mills, Andrew Myers and to all at Elmore Road for their support and encouragement. I benefited greatly from the lively discussion of all who attended my extra-mural class in prehistory at Matlock, with special thanks to Martin Wildgoose for his work on painted hand impressions.

I am grateful to the following individuals for their valuable comments on relevant parts of the book: R. Arambourou, P. Bahn, D. Fine, A. Hooper, A. Marshack, G. McHargue, Dr. P. Mellars, Dr. L. Pales, Dr. K. F. Pardoe, A. Sieveking, A. Stevens and my parents.

The research facilities granted to me at the University Library, Sheffield, the University Library, Cambridge, the Haddon Library, Cambridge, the Institute of Archaeology Library, London, and the Department of Mathematics and Statistics, City of London Polytechnic, are gratefully acknowledged.

1: The Threshold

An Ancient Mystery

When and how did we become human?

Many books on the ancient past state that for over 99 percent of the time span of human existence, man has lived by hunting and gathering. The period that has elapsed since the development of farming and the rise of modern industry is indeed a minute proportion of our past, and yet during this period all but a few scattered remnants of hunting peoples have vanished or have been exterminated. Many argue that even these beleaguered survivors of the Australian outback or the Canadian Arctic are so affected by contact with the Western world that their way of life cannot be taken as a reflection of prehistory. How, then, are we to understand and investigate the largest portion of our existence? To what extent did the hunting past shape our faculties, our instincts, and our relationships in the modern world? For how long, in fact, have human beings thought and felt and acted in ways that would be intelligible today, even if they might seem strange and unfamiliar?

No specialist has ready answers to any of these questions because there are enormous gaps in the scientific record of the most ancient past. To begin with, it is almost an impossibility to picture the slowness of human development through the immense period of the 2.5 million years that separate us from the first known toolmakers in East Africa or the 4 or 5 million years that are currently believed to mark the emergence of distinctly humanlike features in skulls and skeletons of the inhabitants of the same region. In the presence of only a few scraps of information, and faced by an imponderable gulf of time, the scientist still wrestles with fundamental questions: was human development continuous, marked by almost imperceptibly slow and gradual change, or did it proceed by a series of leaps and bounds, divided by stagnant periods? At the present time, many experts in prehistory believe that breaks or discontinuities *are* visible in the record, "thresholds" where important technical or biological leaps appear to have taken place. In the future, as new facts accumulate, some of these stages and divisions may prove illusory or to have acquired more significance than they deserved. At the moment, however, they provide essential steps in a framework, enabling the ordinary imagination to grasp the progress of change across millions of years.

Since the science of prehistory began in Europe a little over a

1

century ago, two such developments have attracted unusual attention and have acquired an emotional significance among scholars and interested laymen. One is the appearance of skeletons that, for the first time, are identical in all important physical respects to those of modern man. The second is the earliest known artworks—the creation of the first representational and symbolic objects, paintings, and carvings which have survived the passing of the centuries.

The exact timing of these events has always been a matter of dispute, and so, too, has the idea that they necessarily originated in the same place or for the same reasons. But, in western Europe at least, the almost coincidental appearance of both the first art objects and the first "modern" men has been a particularly striking fact. In many regions of the west, this apparent threshold is now dated during a few centuries following 40,000 B.C. This date happens to fall in the middle of the cold climatic era popularly known as the last Ice Age. The new advances in our understanding of the life led by hunters and gatherers during this Ice Age, particularly the mystery of the leap forward in the middle of the period, form the theme of this book.

The break around 40,000 B.C., which is also marked by an important change in the stone tools made by prehistoric man, has always been used by archaeologists as one of the fundamental divisions of prehistory; scholars speak of the "Middle Paleolithic" period before it and of the "Upper Paleolithic" afterward. *Upper* here means *Later,* because the highest layers deposited at an ancient site are normally the most recent. While the events surrounding the change are themselves a mystery, even greater questions are raised by the periods before and after it. Did these men, who were so similar to ourselves physically and who left us cave paintings of such beauty that they have inspired the great artists of our own time, really think and behave in ways that we can understand, or compare with the hunting peoples of today? And what of the people who came before them? The people of the Middle Paleolithic who, so far as we know, did not regularly practice art, and whose physical differences from us will forever be linked with the term *Neanderthal.* Were they really subhuman monsters?

Before we can begin to answer these questions directly, it is necessary to view the Ice Age in perspective.

A World of Change

It was a period of decisive change not only in Europe but also all over the world. One of the most exciting aspects of recent prehistoric research has been a significant broadening of knowledge about countries previously neglected as "backwaters" by scholars whose attention has been fixed on ancient Europe for too long.

Change came to South Africa, for example, at much the same time as it did in Europe. For many years it was thought that the South African Upper Paleolithic breakthrough—the earliest appearance of "modern" man and his artworks—occurred thousands of years later than it did in Europe. This assumption cast the South African hunters in the constant shadow of their supposedly more inventive and advanced contemporaries in Europe—as slavish imitators, in fact, dependent on ideas and influences transmitted to them from the north.

Many recent discoveries have helped to shatter this myth. Among the most remarkable of these was the unearthing in 1969 of stone slabs decorated with animal designs in a cave known as Apollo 11, situated in the Huns Mountains north of the Orange River in southwestern Africa. A total of seven fragments were eventually recovered, bearing ill-defined and stiff-legged animals painted with black and white pigments. One of these crudely outlined creatures seems to display a pair of human back legs added at some stage after the original animal legs had faded.

The Apollo 11 paintings are not remarkable for their artistic quality but for their age, which was determined by a meticulous examination of the particular level at which they were found in the cave. There is a strong likelihood that the slabs were painted between 27,500 and 25,500 years ago, which makes them easily the oldest dated works of art yet found in Africa. The peculiar style of the paintings suggests that the artists were in no way influenced by traditions or ideas from western Europe, where the first dated carvings and paintings are probably 6,000 or 7,000 years older than this. The originality of at least one community of South African hunters seems clear. Also, a number of important human fossil finds, although not so firmly dated as the Apollo 11 paintings, suggest that "modern" man was present in South Africa about 40,000 years ago —roughly the same time as in Europe. The notion of South Africa as a prehistoric backwater is therefore obviously mistaken.

An equally remarkable series of discoveries has transformed old ideas about the first peopling of the Australian continent also and has pushed this event far back in time. We know that a hunter of essentially modern appearance inhabited a cave site in Borneo (Niah Cave) around 40,000 B.C. At least one community had installed itself in the highlands of New Guinea some 15,000 years after this, when island-hopping in light water-craft among the Indonesian islands must have been commonplace. By 20,000 B.C. there were many locations on the coastal fringe of Australia (not only in the north and east but also in the extreme southwest as well) that were being exploited by small groups of prehistoric hunters. Among their pioneering activities at this time, they almost certainly walked into Tasmania across the land bridge that existed before the formation of the Bass Strait. It seems likely that the Ice Age Australians were already indulging in the complex practices of rock painting and engraving, which so vividly express the ancient mythology and rituals of the aborigines. There is firm evidence that around 20,000 B.C. a chamber containing crude clay finger drawings at Koonalda Cave in South Australia was sealed from the outside world.

This accumulation of evidence for the antiquity of man in Australia is remarkable enough, but one big discovery dramatizes the facts. In 1969 a remote desert region in New South Wales proved to have been the setting for the oldest burials so far known on the continent. One of these burials was recovered almost intact from the eroding sands of an ancient camp, where simple stone tools had been made, and fish and shellfish had been consumed. Aquatic life of any kind has long since vanished from the parched wilderness of Lake Mungo, which, during prehistoric times, was fed by the seasonal overflowing of the Lachlan river. Here, beside the now extinct lake, a man of entirely modern physical build had been carefully laid out in a grave some 30,000 years ago. The sand underlying the burial was stained a vivid red from the powdered red ocher that had been sprinkled over the body. Red ocher is a natural mineral substance (also known as iron oxide or hematite), and was the coloring material most readily available to many prehistoric peoples; it was used extensively by the Ice Age cave painters of western Europe. Here, on the other side of the world at Lake Mungo, the pigment was used in the burial rites conducted over the body of a prehistoric hunter at a period which would once have seemed impossibly early to Australian scholars. The evidence of the burials implies that the first colonizing of the continent must date back to somewhere around 40,000 B.C.—in other words, to the time when people of physically modern type were also making their appearance over wide areas of Africa and Europe.

There is a fascinating postscript to the Lake Mungo story. Subsequent finds in the southeast showed that not all the early inhabitants of Australia belonged to this advanced type. At a place called Kow Swamp, the remains of an extraordinarily primitive population were discovered, dating about 20,000 years *after* the Lake Mungo burials. We have no idea what the relationship between these two groups may have been, although the contrast in external appearance must surely have been striking. It is simply unknown whether the archaic Kow Swamp peoples were exterminated or died off, or whether some degree of interbreeding is one of the many factors responsible for the present-day physique of the Australian Aborigine.

Today it can hardly be denied that man was in both North and South America long before 12,000 B.C., for more than fifty earlier

The oldest dated art in Australia: crude carvings at Koonalda, South Australia, associated with flint quarries that are at least 20,000 years old.

dates have been calculated from the debris unearthed at excavated sites. However, the number of these pioneers always seems to have been low, and their living sites few and far between, while there are not many human fossils available to help pinpoint their identity. One recent summary of the finds suggests that there were four major phases of colonization and development, beginning perhaps 40,000 or more years ago, although the earliest stages are very poorly documented.

Where did these men come from? With each of the three subsequent waves of colonizing, the dates clearly suggest the passing of new tools, technical ideas, and perhaps people as well, in a slow progress from north to south. It is obvious that these migrants must have crossed the Bering Strait from the Old World, although far too little information is available on either side of the Pacific to detail the movements and relationships. There is a distinct possibility that the earliest hunters to arrive in the New World were not yet of a completely modern physical type.

These significant movements of population were obviously more complicated events than a brief summary can convey. The essential point is that modern research shows how important the Ice Age was for the dispersal of people physically identical to ourselves to practically every continent. The growth of international study and scholarship in this field has been a vital development. What possible excuse can there be, then, for a book such as this one, which retells the story of Ice Age hunters and cave artists in a narrow region of western Europe?

The answer to this question is that the peoples of 40,000 years ago did not form one great colonizing army that swept across the world, but possessed a wide range of cultural and physical identities. An attempt to describe them all, region by region, would result in a textbook cluttered with technical names and descriptions, a catalog of ancient peoples rather than an in-depth portrait. We surely wish to know, as vividly as possible, what it was like to be alive in the last Ice Age, and for this purpose a narrowing of focus is essential.

Fact and Imagination

Prehistoric hunters and their environments have been carefully researched in many countries, and distinguished works have recently appeared, devoted to the Ice Age inhabitants of Britain and western Russia. Yet it is in France more than anywhere else that these studies have been carried to a high level of refinement. This is the result of many factors, including government financing of research, a long tradition of academic expertise, and, above all, the concentration of well-preserved prehistoric deposits protected by the innumerable cave mouths and overhanging cliffs of the southwest. If we seek as detailed a picture as possible of the Ice Age world, it is only natural that we turn to France, as generations of students and scholars have done.

The record is not always an honorable one. The great majority of prehistoric objects on display in museums was not carefully excavated by conscientious scientists but was amassed during the later part of the nineteenth century by private collectors whose methods were often unscrupulous. The threat to archaeological sites at that time came from many sources, as one typical story will illustrate.

The Catholic center of pilgrimage at Lourdes in the Pyrenees happens to be situated close to a cave known as Les Espélugues, which contained late Ice Age tools and superb engraved bones and sculptures. This deposit was explored chaotically by a minimum of seven separate diggers between 1860 and 1885, so it is now impossible to reconstruct the sequence of layers because of all the holes burrowed through them. The magnificent art objects were not found until around 1889, with the activities of the last investigator of the site, Léon Nelli, and he actually uncovered them *outside* the cave. This was because the Fathers of the Cave of the Immaculate Conception had decided to remove much of the deposit to make the interior prettier and to create room for two statues. Some of the earth was taken to a nearby garden and mixed in with ordinary soil; some was used to build an embankment beside a path. It was from this excavated embankment that Nelli retrieved one of the great collections of Ice Age art objects, tools, and ornaments. The objects that had been mixed in with the garden soil were irrevocably lost.

There are many famous sites in southern France with histories as haphazard as this. Nevertheless, it was the curiosity of enlightened collectors in these same regions that first led to the realization of the great antiquity of Ice Age tools and engravings—conventionally dismissed as "Celtic" during the early part of the century. This was due mainly to the work of Edouard Lartet, a former magistrate whose antiquarian activities in the foothills of the Pyrenees led him to suspect the presence of man there long before the period traditionally thought of as "Celtic." His researches, however, began in earnest only when he made friends with a widely traveled English merchant-banker, Henry Christy, who, among other distinctions, introduced Turkish towels into Europe at the Great Exhibition of 1851 held at the Crystal Palace. The displays of Indian and African villages at the Exhibition, which included living inhabitants imported for the occasion, fired him with a passion for the primitive and the exotic.

An early photograph of collectors at work digging holes at Laugerie Basse, near Les Eyzies in southwest France.

A view of the River Vézère, with the cliffs of Les Eyzies in the background, center of the earliest discoveries of prehistoric man and his art.

The encounter with Lartet led to joint explorations, financed by Christy, in the cave sites of the Dordogne region in the early 1860s. It is difficult to imagine the astonishment that these early pioneers must have felt, delving for the first time into some of the richest prehistoric sites in Europe. A critical moment came in May 1864, when, in the presence of several scientists, a realistic engraving of the mammoth, a species extinct since the Pleistocene, was unearthed at the site of La Madeleine. The fact that man had expressed artistic impulses with feeling and accuracy at a very remote age could no longer be doubted.

Despite this dramatic demonstration of the existence of Ice Age art, the first cave paintings and engravings were, until the turn of the century, widely regarded as curiosities or forgeries. The fascinating story of their discovery and authentication has been told many times in popular books. How, in 1879, a Spanish landowner named Marcelino de Sautuola found the magnificent decorated ceiling of Altamira on his own property; how his descriptions of the great painted bisons were at first ignored or ridiculed by most scholars; and how, over twenty years later, a mounting list of discoveries in southwest

Left, the Abbé Breuil (wearing cassock) photographed at El Castillo, northern Spain, in July 1909, with his patron the Prince of Monaco (sitting at right). Right, the pioneer prehistorian Édouard Piette.

France finally overturned skepticism. Among the pioneering advocates of cave art were two great prehistorians, Édouard Piette and the Abbé Henri Breuil, who made essential contributions to the early study of Paleolithic art and archaeology. It is to Breuil in particular that we owe our knowledge of the basic framework of the western Ice Age cultures. Indeed, his copies of cave paintings and engravings, to which he devoted much of his life, have never been surpassed in beauty and accuracy.

This is not a storybook of archaeological discovery. However, it should be obvious that these earliest and most celebrated finds, concentrated as they were in southern France and northern Spain, profoundly influenced general notions about prehistoric hunters. For instance, popular theories about the first appearance of modern man are still deeply biased and colored by discoveries made in France during the early years of this century. We need to understand how these ideas arose in order to appreciate how much recent research has transformed the picture.

The cave paintings of northern Spain and southern France remain the most impressive and intriguing expressions of the Ice Age mind.

A map of decorated caves in western Europe with the names of a few notable or outlying sites. The broken line encloses caves decorated in the distinctive "Mediterranean Style."

As described in Chapter 10, the understanding of the meaning of cave art has not advanced far in fifty years, for no single modern rationalization will match the elusive, diverse qualities of the original artworks. On the other hand, our knowledge of the range of cave-decorating practices and of different forms of art in other regions of Europe grows daily. There are now some 200 painted caves known in southern France and northern Spain. Several new discoveries in the remote Spanish Basque country appear to be closing the gap between these two main regional groups which, in any case, certainly share many correspondences of style and theme.

But there are also outliers located far away from the major centers of activity—the engravings in the cave of Gouy near Rouen, for example, almost at the present shores of the English Channel. Three crude engravings on bone were recovered from cave entrances at Creswell Crags in the Derbyshire county of central England, dating perhaps as far back as 13,000 B.C. There is no reason why cave artists should not have left paintings there or elsewhere in England, although no such traces have been found. Perhaps the sharper climate of the north has obliterated ancient cave walls through the natural forces of erosion more effectively than farther south.

In some regions of Europe where caves were absent—such as Moravia and the Ukraine—rich traditions of engraving and sculpting in bone and ivory flourished. A detailed knowledge of these carving traditions is just beginning to be acquired by scholars in the west. And there are surprises in store for those interested in such regions: for example, the isolated painted cave discovered at Kapova in the Ural Mountains is, so far, the only one of its kind in the whole of eastern Europe. A number of animals painted in red ocher were identified in the Kapova cave, although the style is quite different from those of the west.

There is one final major grouping of Ice Age cave art and engraved objects. It extends around the shores of the northwest Mediterranean and has a character all its own. This "Mediterranean style" of art, which seems to have been little influenced by the master artists of France and Spain, often features simple, stark animal representations together with quite elaborate geometrical designs. There are important caves decorated in the Mediterranean style in southeast Spain, the Ardèche canyons in southern France, the heel of Italy, and in Sicily.

What do these different patterns of artistic activity mean? Were there many different, unconnected ideas? Or was the same code of values—the same system of magic, perhaps—understood and communicated across wide areas, even though it was expressed in varying styles? The question of exactly how artistic ideas could have been transmitted is one of the most fascinating aspects of Paleolithic art.

We cannot begin to suggest answers without forming a general picture of the Ice Age in particular regions: the climate, landscape, and plant life and their effect on human and animal populations. In reconstructing such details, the prehistorian now has indispensable assistance from experts in other branches of the natural sciences. Various dating methods allow ages to be calculated directly from excavated material (for example, by radiocarbon and archaeomagnetic techniques). Less precise dates also can be suggested by careful comparison of the flora and fauna from one period to the next. Such environmental information, whether it comes from the study of preserved animal bones, pollen, soils, snail shells, or beetle wing cases, is now as vital to the archaeologist as are the traditional objects of interest—stone tools and human bones. The combined results of this research are fascinating, although a detailed explanation of each of these techniques would make for tedious reading. (For this reason, brief notes on the more common methods of analysis appear at the end of the book.)

Another transformation of the archaeologist's outlook has taken place outside the laboratory in the approach to digging at the site itself. The early prehistorians of the last century were understandably concerned with establishing the order and sequence of their materials, since no means of exact dating existed to help them. They often sank narrow pits down through the deposit with the sole object of establishing which tools came from which layer, thereby forming a picture of technical changes from one age to the next. These studies are still important but leave unanswered many questions about the kind of society to which the toolmakers belonged. Much broader information is obtained today, not just by digging downward, but by digging outward, or horizontally, at the same time. If the conditions of preservation are favorable, the archaeologist traces and follows the original floor on which the prehistoric inhabitants walked, hoping to expose the remains of hearths and wooden structures. Each layer is successively peeled off over as wide an area as possible. The patterns of flint and bone debris scattered across each floor can give important clues to the particular types of activities that were once carried on there.

The change from narrow pits to broad exposures was an innovation not of the French but of the Russians. In the early years of this century, Russian archaeologists first began encountering thick concentrations of animal bones on the terraces of such great Ukrainian rivers as the Dniester and the Don. Pits sunk here and there could make no sense of the dense bone heaps, but as soon as the remains were opened up horizontally over a broad area, the picture came clear. The bones once formed the foundations and frameworks of houses built when wood was scarce and the shelter of caves unavailable. Since the late 1920s, such bone houses have been found in considerable numbers, often clustered together in little "villages" of four or five houses in the fertile valleys of the Ukraine. The same method of construction has appeared as far west as Kracow, in Poland, where recent excavations in the city center revealed three rings of mammoth bones exactly similar to those in Russia and dating to about 20,000 years ago.

Some of these bone buildings were remarkable structures in their time. One of the most intricate was discovered in 1965 at a place called Mezherich, near Kiev in the Ukraine. A farmer, digging his cellar, almost two meters below ground level, struck the massive lower jaw of a mammoth with his spade. The jawbone was upside down, and had been inserted into the bottom of another jaw like a child's building brick. In fact, as subsequent excavation showed, a complete ring of these inverted interlocking jaws formed the solid base of a roughly circular hut four or five meters across. About three dozen huge, curving mammoth tusks had been used as arching supports for the roof and for the porch, some of them still left in their

View of mammoth bone ruins excavated at Mezin in the Russian Ukraine.

sockets in the skulls. Separate lengths of tusks were even linked in places by a hollow sleeve of ivory that fitted over the join. It has been estimated that the total of bones incorporated in the structure must have belonged to a minimum of ninety-five mammoths. This need not be a measure of some prodigious hunting feat, since gnawing marks of carnivores suggest that many of them were scavenged. However, the task of dragging the enormous skulls across country should not be underestimated since a small one weighed about one hundred kilograms. It is likely that this extremely solid framework, when completed, was covered with hides just like the skin-and-whalebone huts built by Siberian coastal hunters during the nineteenth century.

Inside the Mezherich building, there were some remarkable finds: amber ornaments and fossil shells, transported an estimated 350 to 500 kilometers from their source, and the remains of one of the earliest percussion instruments ever found. The "drum" consisted of a mammoth skull set at the entrance porch and painted with a pattern of red ocher dots and lines. The top of this skull bears depressions where it seems to have been beaten by "drumsticks," the animal long bones that were found to bear corresponding damage on their ends. It is possible that the building may have served some ritual or communal function at which the mammoth bone rhythms were beaten out, although many Ukrainian huts of a similar size seem to have been ordinary living places.

In the far west, there was simply no need for such solid structures because shelter was provided by limestone overhangs and cave mouths. Nevertheless, the same principles of excavation pioneered in Russia have revealed traces of "interior design" at a number of cave sites, such as the remains of wooden partitions and lean-to structures that would have helped to exclude cold and damp. In addition, temporary open-air camps are being discovered in increasing numbers in western Europe. Several of these appear to be summertime halts occupied by small groups of hunters who may have joined up with larger groups to live in the shelter of caves during the winter. At the most thoroughly investigated of all these camps, situated at Pincevent in the Seine valley, a meticulous attention to detail allowed the excavator to reconstruct the outlines of three light skin tents occupied by a small summer band of reindeer hunters. Although no wooden or bone foundations had survived, the shape of the structures was revealed partly by the pattern of flint flakes struck off by toolmakers inside the tents. The paths traced by some of the flakes had clearly been interrupted by the hide walls.

As these few examples illustrate, there have been enormous advances in the practical methods of extracting information from an Ice Age site. Both in the field and in the laboratory, the expert can reconstruct aspects of the environment and of human behavior in the

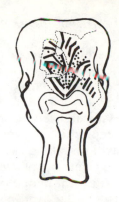

A model of the hut, built of mammoth bones and tusks, that was excavated at Mezherich in the Ukraine. At the right-hand side of the hut entrance was set this mammoth skull with a red-painted design, which seems to have been used as a drum.

kind of detail that was undreamed of by scholars of an earlier generation. Despite all this scientific progress, however, there are still many gaps in the record and many areas of inquiry where the expert is unable to help us, even with the most refined technology available. Prehistory is not just the sum of bits and pieces of technical information. Clearly, we want to know more than this, to build for ourselves some over-all picture of the ancient past. The question of how to proceed beyond the basic factual evidence has perplexed prehistorians for a long time. Should we just use our modern imagination? Or is it reasonable to take the records of primitive peoples of the recent past as a guide to the vanished features of prehistory?

There have been, and are, wide differences of opinion on this issue. Some scholars have flatly denied the relevance of the Australian Aborigines or the Eskimos to an understanding of Paleolithic hunters. One of the most distinguished French investigators of cave art, André Leroi-Gourhan, believed that the evidence should speak for itself:

> To take what is known about prehistory and cast about for parallels in the life of present-day peoples does not throw light on the behaviour of prehistoric man.[1]

At the opposite extreme, many early commentators took puzzling objects recovered from Paleolithic sites or the strange signs painted on cave walls and interpreted them with absurd ease by picking on some isolated practice of modern primitives that just happened to fit. Other possible explanations, which might fit just as well or even better, were ignored in such a literal "translation" of the present into the past.

Many archaeologists today believe that there should be room for a compromise between these two attitudes. In practice, it is difficult to avoid thinking of the behavior of modern peoples when seeking an explanation of some archaeological fact. In the course of this book, a wide range of ethnographic evidence is described, not because the Ice Age hunters are assumed to have been just like the Bushmen or the American northwest coast Indians. Instead, recent sources of information are used simply to provide very general ideas about the way in which Paleolithic people *may* have treated the reindeer or exchanged art objects. These ideas can then be accepted or rejected by comparison with the actual archaeological material (if enough of it exists to swing the argument one way or the other). In this way we can hope to paint a vivid over-all picture of Ice Age life without making rash assumptions.

Many writers and scholars have tended either to romanticize or to brutalize the hunting past. In the chapters that follow, a few of these moral attitudes are briefly exposed, and tested against the scientific information. Why should Neanderthal man have acquired such a persistent and unsavory reputation as a subhuman monster? Was the Paleolithic really a golden age of contentment made possible by unlimited resources and scant human needs? It is suggested that most moral judgments imposed on the Ice Age fail to match the character and complexity of the facts now available. The strangeness and intricacy of the evidence far exceeds simple notions of prehistoric hunters as instinctive brutes or noble savages. Most preconceived ideas are simply inadequate for coming to terms with the secrets of the Ice Age.

Australian Aborigine hunters photographed recently in Arnhem Land, Northern Territory.

2: Three Million Years

The Rhythms of Ice

There have been many ice ages, and many of these occurred before man. The idea that modern glaciers once extended far beyond their present-day limits was first suggested by a number of observers in the Swiss Alps at the beginning of the nineteenth century. There was much physical evidence that demanded explanation in terms of intense glacial activity: huge erratic boulders stranded far from their original source, heaps of stony debris, moraines deposited far from the melting margins of recent glaciers, and—most conspicuous of all —rock outcrops that had been polished and scored by the movement of overlying ice. In 1837, a young Swiss zoologist, Louis Agassiz, profiting from earlier observations of these phenomena, suggested that a worldwide drop in temperature at some remote epoch in the

The edge of the polar ice sheet, photographed in southwest Greenland.

past had been responsible for the spread of the glaciers. It was Agassiz who coined the term *Eiszeit,* or *Ice Age,* to refer to this era, and in his heroic vision of the period, he imagined that the ice had once stretched from the North Pole to the shores of the Mediterranean. Within a few years he did indeed find widespread confirmation of his theories, as, for example, when he landed in the New World in 1846. Disembarking at Halifax, Agassiz wrote that he

> sprang on shore and started at a brisk pace for the heights above the landing. On the first undisturbed ground, after leaving the town, I was met by the familiar signs, the polished surfaces, the furrows and the striations, so well known in the Old World. . . .[1]

In 1961, a team of French and Algerian geologists prospecting for oil in the central Sahara recognized these same signs of glacial activity, the scarred and polished bedrock overlain by deposits and debris so ancient that they had become cemented into solid rock. The presence of fossil animals in these deposits, and the geological position of the rocks themselves, permitted an estimate of some 450 million years since the period when ice stretched over the Sahara.

Similar indications suggest the activity of other remote ice ages in many countries of the world: hardened glacial deposits in Canada dated over 2 billion years ago, evidence of ice sheets in Russia and China dated over 950 million years ago, and so on. Indeed, the pattern formed by these scattered points in time has led some investigators to propose an immense recurring cycle of warm and cold that is roughly 200 or 300 million years in length.

This is the grandest of the rhythms that appear to have regulated the earth's climatic history, rhythms for which geologists and physicists still have not agreed on any single or comprehensive explanation. For example, in accounting for these very ancient periods of glaciation, it is tempting to imagine that the slow drift of the earth's continents may have shifted large land masses to the present polar regions, where the presence of high mountain ranges would have encouraged the growth of huge glaciers. In addition, the solar heat reflected away from the earth by these great white masses would have provoked a further drop in temperature, and so an ice age would begin. This theory, however, is undermined by the same methods that enabled geologists to prove the theory of drifting continents. Such studies show that the present configuration of the polar regions was in fact established many millions of years before the last major burst of glacial activity began. This is the ice age that is most thoroughly known to geologists and in which our present-day climate may only represent a temporary lull. If so, then we are still experiencing the effects of the last great ice age, the Pleistocene.

Because of its proximity to us, there are obviously more traces of the Pleistocene epoch for geologists to study than there are for remoter ice ages. However, there are other reasons for the intense research devoted to this period, including the emergence of many present-day mammal species, such as the horse, camel, elephant, and of course man himself. Traditionally, the length of the Pleistocene has been put at 2 million years, but recent research suggests that climatic deterioration had set in 3 million years ago, or perhaps even considerably earlier. This great time span was not an age of incessant cold, an unbroken and monotonous "deep freeze," for here we encounter the second baffling rhythm to which the earth's climate seems to have fluctuated.

Within the Pleistocene age, alternating phases of cold and mild climate appear to have been accompanied by a vast range of natural responses. The cold temperatures of course encouraged the spread of glaciers from the poles and from mountain ranges, but as if to compensate for land lost under the ice sheets, huge expanses of coastal plain were exposed. The periodic lowering of sea levels by as much as one hundred meters or more was the result of the huge volumes of water locked up in the ice sheets. Thus innumerable traces of prehistoric man, including all of his coastal settlements in many regions of western Europe, now lie far out under the ocean.

THE DIFFERENT LEVELS OF CLIMATIC CHANGE

	Appropriate timescale unit	
1. Minor fluctuations in meteorological records	10 years	Irregular fluctuations which seem to operate over intervals of about 25–100 years.
2. Post-glacial and historic changes	100 years	Changes over intervals of about 250–1,000 years such as the "Little Ice Age" of 17–18th century Europe.
3. Glacial periods	10,000 years	"Ice ages" as the term is used in this book, which is mainly about the last ("Würm") Ice Age—some 50–60,000 years long.
4. Minor geological changes	1 million years	The duration of major phases of glacial activity as a whole.
5. Major geological changes	100 million years	The spacing of these major phases of activity, apparently about 200–300 million years apart.

Main Sources of Evidence

1. Meteorological records and those of river and lake levels, crop yields, tree rings, etc.
2. Historic and archaeological records, tree rings, sediments in lakes and oceans, pollen, radiocarbon dating.
3. Fauna and flora characteristic of different environments, as reconstructed from pollen, animal bones, etc. Ocean bed sediment samples. Geological evidence.
4, 5. Geological evidence, fossil flora and fauna, various radioactive dating methods applied to rocks.

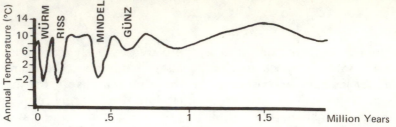

A summary of the evidence for temperature changes in central Europe during the Pleistocene.

During milder intervals, sometimes more favorable than the climate of today, the waters rose to create beaches high above the present sea level, and by comparison of the heights of these beaches, experts can recognize (not always with total confidence) the mark of a particular warm phase. At the same time, the land, freed of its burden of ice, rose up in some regions by hundreds of meters. All of these physical changes had profound effects on the zones of vegetation, which swung back and forth through as much as 10 degrees of latitude, and on the corresponding habitats of animals and man.

The scale of these environmental changes certainly matches that of Agassiz' grand Ice Age vision, but we now know that the Pleistocene glaciers never reached the Mediterranean as he believed. At their maximum extent many hundred thousand years ago, the world's ice sheets covered about three times the area that they presently occupy. In Europe and Asia, great systems of glaciers arose from centers in the Alps, Siberia, and Scandinavia, building up to frozen masses that in the north may have been 2,000 or 3,000 meters in thickness. Even during the most severe phases, the Alpine ice never joined the Scandinavian, so a corridor of land running north of the Alps always existed. At one stage, this corridor can only have

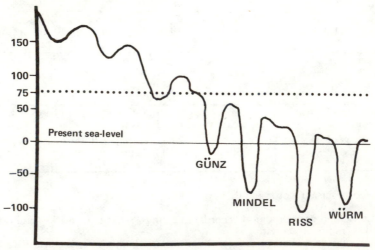

Sea-level changes during the Pleistocene. The dotted line shows where the level would be if the modern ice caps were melted.

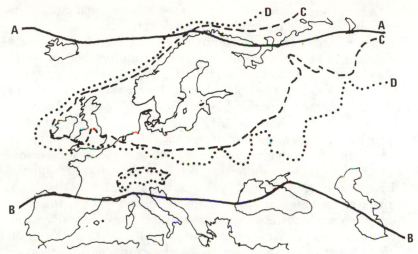

Ice Age conditions in Europe.
A. The position of polar timberline in present-day Europe.
B. The position of timberline at the most severe stage of the Würm Ice Age.
C. The limits of glacial debris deposited during the Würm Ice Age.
D. The limits of glacial debris deposited during the Riss and Mindel ice ages.

been a few hundred kilometers wide, as the whole of northern Europe, including Britain, the North Sea, and Scandinavia, lay deeply buried under ice, but this was exceptional. At other times during the recent episodes of the Pleistocene, large areas of the north were exposed, and under the most adverse cold it was almost certainly possible to walk across land into southern England at a favorable season of the year.

The fluctuations of warm and cold that left so many marks on the face of Pleistocene Europe are the focus of intense study and disagreement. Only four such waxings and wanings of the ice sheets were identified at the turn of the century, based again on observations in the Swiss Alps, where the four ages were given the classic names, Günz, Mindel, Riss, and Würm, which have remained popular despite the fact that today an estimate of eight major warm–cold cycles during the same period of time would probably be conservative. As techniques of detection have become increasingly refined, the number of glacial advances and retreats that crept across Europe has grown. Although the old names are hopelessly inadequate to refer to such a complicated sequence, they are still widely used in western Europe (usually subdivided with the help of numerals to refer to the cold spells, such as Würm I, Würm II, Würm III, and so on). But this system itself applies only to the west: in other parts of the world, observations based on the Alps are irrelevant to the to-and-fro of climate recorded there, and so a host of different names and labels have been spawned. One of the problems that geologists and archaeologists face is simply that of description.

Whatever we may choose to call them, the fluctuations of climate do not appear to be entirely random. It is intriguing that the cycle of cold and warm appears to repeat itself approximately once every 100,000 years, and that the cold phase is by far the longer part of the cycle; indeed, the temperate interval between ice ages in general seems to last for only about 10,000 years.

Scientists have expended enormous ingenuity in attempting to account for this fundamental rhythm. Among the more plausible theories are the slight irregularities in the earth's orbit around the sun, changes in solar radiation, cycles of volcanic activity that blow heat-reflecting dust high into the atmosphere, and so on. However, no single theory adequately explains both the brief, episodic variations in climate and the immense time gaps of 200 to 300 million years that appear to separate the major phases of glacial activity.

Nevertheless, the apparent regularities in the spacing of ice ages have recently inspired some sensational doom-laden prophecies. If the average length of a warm interval between ice ages (an "interglacial") is really on the order of 10,000 years, it may be that we have outstayed our welcome in the present temperate era and that a new ice age is imminent. But, before the specter of ice sheets towering over Manhattan is taken seriously, our ignorance of the precise forces and the timing involved in a major advance of the glaciers should be kept firmly in mind. The problem is likely to remain one of distinguishing a possible large-scale deterioration of climate from some short-lived warm spell or cold snap. One of these brief episodes ensured that eleventh-century southern English wine was the envy of the French wine producers, while another, the so-called "Little Ice Age" which seems to have reached its peak during the eighteenth century, helped to encourage the widespread desertion of upland farms in northern Britain and Scandinavia. In the short run, the prospect of increasing warmth because of the greenhouse effect of fossil fuel burning is a far more serious and imminent prospect than that of glaciation. It could be that the next major climatic phase will be largely of our own making.

This book is not about prophecy, but about the past. The "Ice Age" of the title could be misleading; it does not refer to the vast geological scale of changes from one 300 million-year epoch to the next, nor even to the developments which took place throughout the 2 million years of our own era, the Pleistocene. Instead, this book is about just one single age of glaciation, the final burst of cold that overtook Europe before the present period of climate began in roughly 9000 B.C. There are many figures quoted for the starting point of this last glaciation; an estimate of about 70,000 B.C. would be conventional. This is the period known within the classic Alpine system as the Würm. Elsewhere the last glacial advance has its appropriate name: in America, the Wisconsin; in Britain, the Devensian; in western Russia, the Valdai, and so on. Throughout this book, the term *Ice Age* is used as a convenient shorthand to refer only to this restricted period of the Würm.

Here, then, is a mere 60,000 years or so selected from the unimaginable time spans of the geological record. Why is this period so critical for an understanding of human development? Why this Ice Age and not a dozen earlier ones? We will try to answer these questions in the following chapters.

The Sluggish Ascent

The last Ice Age was a period when, compared to the hundreds of thousands of years and the numerous glacial episodes preceding it, human development seems to have quickened and diversified. This time of change is best symbolized by the division mentioned earlier between the Middle and the Upper Paleolithic—the break or transition after 40,000 B.C. that ushered in physically modern man and a wide range of artistic expression. Although this development does indeed claim our attention, it should not be supposed that in our prehistoric past this was the only significant threshold or that the human mind was, in its creative workings, a "blank sheet" before this time. Just as we have glimpsed the last Ice Age set in a vast perspective of geological and climatic change, so it is important to grasp something of the huge time spans underlying our mental and physical development.

Even the most primitive cultural remains that have survived are, from an evolutionary point of view, already far advanced in the direction of humanity. At the earliest sites, currently under intensive study in the arid Rift valley regions of Kenya and Ethiopia, we encounter evidences of behavior far removed from the normal activities of the great apes.

Between 2 and 3 million years ago, the shores of the ancient lakes in these regions were inhabited by a variety of upright two-legged manlike creatures (or *hominids*), with a brain capacity roughly half that of the modern average. These hominids regularly smashed apart hard volcanic pebbles and other stone materials to create crude but effective chopping and cutting tools. At some of the East African sites, it is known that suitable stone must have been transported a considerable distance (in the case of tools found at one site near Lake Turkana, surrounding the butchered carcass of a hippopotamus, the material must have been carried at least three kilometers). The tools have been found in clusters, sometimes isolated from any other finds, but occasionally associated with the butchered remains of several different species of animal. It is likely, although there is no positive evidence to support it, that the hominids carried lumps of stone, joints of meat, and perhaps vegetable foods as well, in primitive containers of bark or hide.

Transporting food and tools back to a "home base," which implies the sharing of food with others not directly engaged in hunting or scavenging (the females?), is a distinctly human, as opposed to primate, pattern of behavior. "Tools" of twigs and grass stems, for instance, are occasionally used by chimpanzees in the wild, but their social arrangements never include the habitual sharing of food at a fixed base.

There is no space here to discuss the full significance of these discoveries, which have been the subject of excellent popular accounts by the archaeologists involved, notably Richard Leakey and

Richard Leakey applying hardening fluid to a fossil jawbone found in a dry streambed at Lake Turkana, Kenya, center of research for the earliest remains of man.

Glyn Isaac. The essential point is that by the time archaeological remains of stone and bone are present for study, at around 2 million years, they already point to behavior closer to that of modern hunting and gathering peoples than that of any colonies of primates observed in the wild.

This conclusion is reinforced by intriguing scraps of evidence that suggest the existence of rudimentary creative faculties among the hominids. While individual chimpanzees and gorillas raised exclusively in human company have become famous for their ability to "speak" in sign language and to paint pictures, no activity recorded among primates in the wild could have resulted, for instance, in the crude stone wall uncovered at Olduvai Gorge by Louis and Mary Leakey. This arrangement of large stones, set in a rough semicircle and dating to about 1.8 million years ago, may have served as the foundation of a hut or windbreak. At similar sites in Olduvai Gorge, pebbles composed of green lava were carried in to the living area of the hominids, who were perhaps attracted by the distinctive color. One of the stone cobbles examined by the Leakeys revealed more than just the haphazard workmanship characteristic of the earliest tools. It exhibited grooves and pockmarks produced by another stone tool, including a line of four symmetrical pits. According to Mary Leakey,

> it seems unlikely that it could have served as a tool or for any practical purpose.[2]

An even more remarkable discovery was the presence of a fragment of red ocher on one of the Olduvai campsite floors. However crude their sensibilities were, the Olduvai hominids may have responded to color.

The excitement of recent discoveries in East Africa tends to obscure one important fact: the earliest human record is not one of rapid innovation and ingenuity but of almost inconceivable stagnation and conservatism. Certain features of the early hominid skulls, notably the form of the teeth and jaws, remained essentially unchanged for millions of years. It is particularly striking that brain capacity seems to have stayed fairly constant at around 600 to 800 cubic centimeters (a little over half the average modern capacity) for a period approaching 2 million years in length.

This apparent lack of intellectual progress may be reflected in the absence of any important additions to the hominids' range of tools during the same period. Indeed, Mary Leakey has demonstrated that for one interval of 500,000 years, which she has studied closely at Olduvai Gorge, literally no change is registered in the character of either the sites or the stone tools found there. Only later, toward 1 million years ago, did recognizable and repeated "standard" tool

The earliest known tools consist of simple flaked pebbles, like these found at Olduvai Gorge.

forms begin to prevail over the rough and ready use of whatever stone fragments lay to hand. In technical matters and for an immense period of time, it seems that our earliest ancestors remained opportunists rather than craftsmen.

When change eventually did come, it seems to have occurred with relative swiftness and to have affected not only standards of toolmaking but also physical development and the spread of populations outside the African continent. Our understanding of these events is still extremely poor, but the period around 1 million B.C. seems to represent an important threshold of change, before renewed conservatism gripped the faculties of prehistoric peoples during the era leading up to the last Ice Age.

The earliest indications of change are once again in East Africa, dating to around 1.5 million years ago. At this time, hominids appeared with facial features that in some respects must have resembled earlier fossils. However, they also displayed novel characteristics, such as a narrower face with more pronounced eyebrow ridges and a brain capacity that eventually reached levels of around 1,000 cubic centimeters. At the moment, the background to this physical development is not at all clear, so the evidence of changes in tool forms is of special importance. From this same period come the first crude examples of perhaps the best-known flint implement in prehistory, the hand-ax.

The easiest explanation for the appearance of the hand-axe is that the impulse behind its manufacture lay far back in the minds of the early hominids who created simple pebble chopping tools. Not content with a single crude cutting edge, they began to extend the working of the choppers around the sides of the pebble. Eventually, it was customary to fashion the entire surface of the stone, although there was a considerable variety in both the shape and the quality of the finished tool. The importance of this development is that probably for the first time a preconceived shape was regularly being sought and created in stone, rather than a mere sharp edge that could be acquired by fairly aimless smashing. Despite the significance of this advance, we have very little idea of what hand-axes were used for, other than that (as the name implies) they were probably gripped in the hand and used for a variety of chopping and cutting purposes. The size and sturdiness of some hand-axes could well reflect the difficulty of dismembering large game animals.

The earliest hand-axes do not in themselves seem to have constituted a prehistoric industrial revolution. Indeed, quite conventional collections of pebble tools—both with and without accompanying hand-axes—continued to be manufactured for at least 500,000 years after its initial appearance. It seems most unlikely, then, that big-browed "invaders," armed with the technological superiority of the hand-axe, dislodged a more primitive race of toolmakers from its traditional lakeside haunts at Olduvai and Turkana. Instead, it is probable that the indigenous hominids of East Africa achieved this physical and technical progress for themselves and in addition began to break away from the simple pattern of hunting or scavenging that they had followed for countless generations. From about 1 million years ago, the stone tools are found not only at the usual lakeside "home bases" but also at smaller sites that may represent temporary halts in the course of hunting expeditions. Caves and springs were chosen for settlement, while coastal sites in North Africa, at present poorly dated, show that marine sources of food were eventually exploited.

Since this book is concerned with the prehistory of Europe, the question of when man first ventured outside Africa is of special interest. There are tantalizing hints that some initial colonizing of the Far East and Europe may have been achieved by makers of pebble tools well before 1 million B.C., although as yet no reliably established dates exist to confirm this possibility.

Some of the strongest evidence for human activity in Europe at such a remote age comes from a cave in the Provence region of southern France known as Vallonet. Here, a very small number of artificially worked pebbles was found, along with the bones of numerous animals, including several extinct varieties. A recent estimate of the age of the Vallonet finds, based mainly on the types of animals represented by the bones, is 900,000 years. Whoever dragged these animal remains into the cave disposed of the major bones by abandoning them against the cave walls. Deer antlers, some of which appear to have been artificially worked, were also imported into the cave. So, too, were fragments of a whale, no doubt discovered in a decomposing state on a nearby beach. It is interesting that no trace of any fires was detected at Vallonet.

If it is true that hominids with a rudimentary chopper technology were present in Europe as long ago as 1 million years, then they were there in very low numbers. It is only at a considerably later date that prehistoric people, manufacturing hand-axes, began to settle in western Europe in earnest. This period of expansion was associated from its earliest beginnings, about 500,000 years ago, with a mastery of fire.

The direction from which the pioneers came is fairly clear: it was from North Africa. Most of the hand-axes are found in lowlying river valleys draining into the Atlantic. The discoveries are thickly concentrated in the Paris Basin; the activities of early collectors in the gravel pits of the Somme valley area ensured that some 20,000 hand-axes had been identified there by the turn of the century. Important groups of sites also are located in southern England, southwest France and Spain. In central Europe and farther to the east, the picture is quite different: many diverse tool traditions, but no sign of the hand-axes. This pattern of finds strongly suggests that the western settlers came across the Strait of Gibraltar either in simple boats or even by swimming. During the periods of most intense cold, when an enormous volume of ocean water was frozen in the ice sheets, it is estimated that the width of the Strait could have narrowed to as little as five or six kilometers.

What was life like for the early Europeans, two or three ice ages ago? We are still ignorant about the great interval that elapsed between the initial peopling of Europe and the final major advance of the glaciers, around 70,000 B.C. Human fossils are frustratingly rare compared to later periods; as one scholar recently commented: "the record is much patchier and the dating quite insecure."[3] A mere handful of sites in the west have given a detailed, vivid picture of the activities of a particular group of prehistoric hunters. The insights gained from two or three excavations have nevertheless been remarkable.

At Terra Amata, for example, we have the earliest huts known in Europe, built of light branches on the beach at Nice, when it was almost thirty meters higher and perhaps 300,000 years older than it is today. We know that one of the huts was rebuilt eleven times on exactly the same foundations, in the course of brief seasonal visits to the site at the end of spring or in early summer. The huts were heated and lightly paved, while inside the inhabitants consumed a varied diet consisting mainly of young deer, but also of elephant, pig, turtle, fish, and oysters. The inhabitants manufactured stone tools and used lumps of red ocher for some unknown purpose. The intimate details recorded at Terra Amata include impressions in the sand of a wooden bowl, of skins that in places carpeted the hut floors, and of the earliest known human footprint, which appears likely to have corresponded to a person about 1.52 meters in height.

To put ourselves in the minds of prehistoric hunters at this remote date is a difficult challenge. Their instincts and beliefs, if indeed they were able to express them articulately, would surely seem strange to us. At the cave of Arago, set steeply above the Roussillon plain not far from Perpignan, human bones repeatedly found their way into the cooking debris discarded by a group of hunters. These bones included many isolated teeth, part of a jaw that probably belonged to a twenty-year-old male, and another jaw from a woman in her forties or early fifties. In 1971, a remarkable discovery was made of a twenty-year-old's skull, lodged upside down in the midst of a dense layer of tools and animal remains. This skull presents us with one of only about three faces that are known from the entire Riss glaciation. The appearance of the Arago head would seem remarkable to us because of its flatness, pronounced brow ridges, and projecting face, but it was less extreme in these respects than several human skulls from Africa dating to the same era. At an early date, it seems that the populations of Europe had already begun to develop physical characteristics distinct from their African forebears.

The Arago skull, discovered in July 1971, near Perpignan in southern France.

 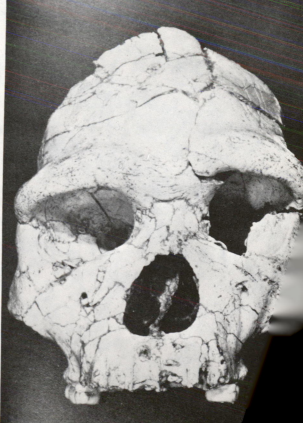

The evidence of the Arago head is not entirely satisfactory because the whole back half of the skull is missing, and was perhaps deliberately removed in order to gain access to the brain. The cave entrance in which this individual was found almost certainly looked out across a plain filled with grazing herds of horse and rhinoceros. The presumed cannibal feast took place among a thick mass of discarded bones, including those of horse, rhino, deer, cattle, ibex, and lion. It seems distinctly improbable that the young person's brain was consumed because of pressing hunger. Instead, it may well be that the activities uncovered at Arago represent the early stirrings of ritual anxiety and belief, however repugnant their character might seem to us.

These symptoms of intellectual advance are interesting, but the record of the Riss Ice Age as a whole is not notable for technological inventiveness. The toolmakers were still practicing a repertoire of only some seven or eight basic types of implement; they seem to have

Part of a plan of the level in which the Arago skull was found, lying upside down among a thick mass of discarded animal bones, tools and other debris.

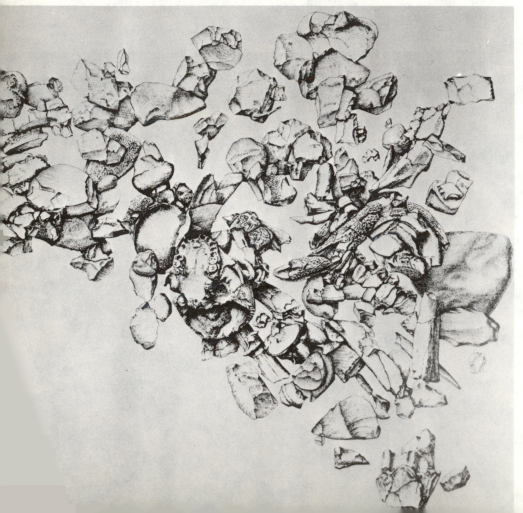

A contrast in the shape of hand-axes excavated at Swanscombe, Kent, England.

been concerned not with the development of new ideas but with the refinement of old ones. The skill and craftsmanship devoted to these traditional forms are often astonishing, far exceeding the demands of a practical working tool. For the first time, preferences of style are clearly apparent in the hand-axes, which vary from heart shapes to much narrower forms and may reflect the tastes of individual craftsmen.

Nor is the evidence for an appreciation of symmetry and texture limited only to the hand-axes. Roughly battered round lumps of stone are present in the earliest hominid tool kits, where they may have served as crude missiles of self-defense hurled against predators, while at a later date, more regular spherical shapes appear. These could have been effective hunting weapons if they had been

attached to long leather thongs and twirled in the air like the bolas
of the Argentinean cowboys and others. Indeed, Louis Leakey himself was fond of demonstrating their effectiveness in bringing down
small African game. On the other hand, as in the case of the hand-
axes, practical demands were not always uppermost in the minds of
the craftsmen. Perfect stone spheres, unnecessarily regular for any
possible use as missiles, were laboriously made from alluring but
intractable materials like quartz. Were these magical or mystical
objects in prehistoric eyes?

The early Europeans may at least have had the capacity for symbolizing their thoughts, even if they rarely gave expression to them in
ways that have survived. The world's first known engraving dates
from the early Riss Ice Age and may be about 200,000 years old.
It appears on the rib bone of an ox that was found during the course
of excavations at Pech de l'Azé, situated in the Dordogne valley of
southwest France. The markings consist of three arclike patterns that
overlap one another, showing that the design was executed in an
orderly sequence. There also are very faint lines and V-shaped marks
that could be the result of natural damage in the ground but also
could be intentional. Was the engraver trying to represent something with these faltering images? Or had he merely discovered the
pleasure of "doodling" for its own sake?

The Pech de l'Azé bone represents an isolated artistic experiment
in an age that otherwise appears to be bereft of innovation. The
contrast with the ice age that followed could not be greater. Already,
by the time a warm interval of climate intervened shortly before
100,000 B.C., there were signs of an important "threshold" of
change. Craftsmen continued to make hand-axes, but now a sophisticated technique of preparing the stone was widely adopted for the
first time; this resulted in a general reduction in size and an improve-
ment in the quality of many tools. At the same time, a remarkable
range of physical types seems to have flourished, and out of this
diversity, there eventually emerged the most distinctive and famous
of all prehistoric Europeans, the Neanderthals. During this warm
interval, too, perhaps the first wind music was heard piping from the
cave entrance of Haua Fteah on the North African coast. The bone
flutes recovered there are the earliest musical instruments known.

The full story of these developments is told in the chapters that
follow. Only one important question remains to be considered here:
why did so much happen during the last Ice Age and not the earlier
ones? What new factors could be responsible for this apparent quick-
ening of the pace of cultural and physical progress? For nearly 500,-
000 years, ancient European populations had been subjected to glacial advances and retreats, sometimes more drastic in their effects on
the climate and landscape than anything that was to be experienced
during the Würm. Yet as we have seen, the human traces recovered

from these earlier ice ages seem to have been the product of an extreme conservatism. This may lead us to suppose that a major impulse for change lay not in the environment but in man himself. In particular, it is often suggested that only when language had fully developed could the pace of human inventiveness accelerate to any important extent. New ideas could be more securely transmitted and shared through speech than through any silent example or gesture.

Unfortunately, it is impossible to know exactly when we became articulate. Recently, a number of ingenious investigations of prehistoric human skulls have been launched in hopes of reconstructing vocal tracts or of identifying speech centers in the brain by the inside "bumps" of the skull. So far, these novel studies have met with contradictory results, and there is no point in prehistoric time at which we can objectively assume the presence of language. However much common sense suggests that Europeans had begun to think and talk like us as early as 100,000 years ago, more experimental work is necessary to prove the point.

Indeed, for over a century, the issue of the physical character and capabilities of early Ice Age people has been clouded by facile assumptions and emotional arguments. The history of this debate, and of recent scientific attempts to dispel the confusion, is considered in an attempt to answer the question: who exactly was Neanderthal man?

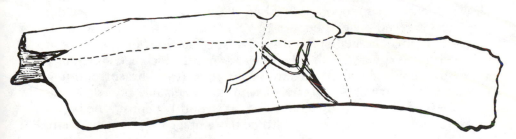

The world's oldest engraving on an ox rib from Pech de l'Azé in the Dordogne region, dating to the early part of the Riss glaciation. There are many other markings on the bone, most of which were probably due to naturally caused damage in the ground. The bone is nearly 17 centimeters long and may date to about 200,000 B.C.

3: In Search of the Neanderthals

The Abominable Caveman

Of all the bewildering terms applied by science to varieties of early man, none is more familiar than Neanderthal, and no other ancestor of the human race has been a more constant target of popular abuse. At the mention of Neanderthal man, we usually picture the club-wielding apeman so often depicted by cartoonists. His image is such a subhuman one that popular writers often propose that surviving colonies of Neanderthals are the fur-covered Yeti or Abominable Snowman. Until about twenty years ago, equally unattractive impressions of Neanderthal man persisted in the works of archaeologists and anthropologists. Since then, the recovery of respect for Neanderthal man and his abilities has been one of the most important consequences of prehistoric research.   :

The invective leveled at Neanderthal man dates back to some of the earliest interpretations of the original discovery. The name itself derives from the Neander Valley, not far from Düsseldorf in Germany, a secluded, winding ravine, bordered on one side by sheer limestone cliffs. During 1856, workmen engaged in building a railway for quarrying operations along the steep southern side of the valley revealed the existence of two caves in the limestone about 18 meters above the bed of the River Düssel. The floor of one of these caverns was covered with a thick layer of mud, which had apparently washed down through a fissure in the rock from the surrounding plateau country and had carried with it the remains of a skeleton. Preserved by chance at the bottom of this muddy deposit, the bones were thought by the workmen to be those of a cave bear, and many of them were thrown away.

The surviving fragments were eventually examined by a local schoolteacher and naturalist, Johann Carl Fuhlrott, but the circumstances of the discovery afforded him no clues to their possible age. Nevertheless, Fuhlrott decided that the thick skull cap, with its projecting eyebrow ridges, must have belonged to a very ancient member of the human race. He then passed on the remains to an anatomist at Bonn University, Dr. Hermann Schaafhausen, who concurred with his judgment. Despite the lack of any actual evidence to date the bones, Schaafhausen assigned them to the period of the Diluvium, the time before the Flood, when extinct animals roamed through the Neander gorge.

It was obvious to all investigators that the unusual features of the

34

Neander bones demanded some special explanation. The skull itself was of a remarkable size and thickness, and its shape was abnormal. It was long and wide, with an insignificant forehead, but it had massive projecting ridges around the eye sockets. The dimensions of the surviving rib and leg bones were similar to those of modern Europeans, but they were remarkably thick, and their attachments indicated powerful muscular development. Schaafhausen and others concluded that these differences could only be explained if the Neanderthal bones were "the most ancient memorial of the early inhabitants of Europe."[1]

Yet as soon as these views were made known in 1857, Neanderthal man had his detractors. Some were reputed men of science who could not conceive of such an extraordinary specimen having any connection with man. One of Schaafhausen's colleagues at Bonn, Professor Mayer, declared that the bones belonged to a Cossack who had crawled up to the cave to die when the Russian army crossed the Rhine in 1814 and who was suffering from rickets. The upper thigh bones were curved inward because the unfortunate Cossack had been accustomed to riding horses from childhood.

A grotesque vision of Neanderthal man, published as the frontispiece to H. F. Osborn's book *Men of the Old Stone Age* in 1915.

An even more vigorous opponent of Schaafhausen's views was the eminent anatomist, Rudolf Virchow. Because of his reputation among scientists throughout Europe, his verdict almost condemned Neanderthal man to oblivion. Clearly, Virchow maintained, the bones belonged to a pathological individual who, although ravaged by rickets and arthritis, had managed to survive into old age. Such advanced years would only have been possible within a settled agricultural community, not in the precarious living conditions of the Ice Age. Moreover, skeletons dating from the Ice Age had recently been found in France, and they were identical to modern man. Virchow was therefore able to proclaim confidently not only that the arthritic old man of Neander had nothing to do with ancient humanity but also that ancient humanity itself had no connection with lower animal forms. Virchow's opinions of the bones were echoed by many of his contemporaries. For example, Professor Gratiolet was convinced that the Neanderthal skull was similar in every respect to "an idiot's head of the present day," which he is reported to have exhibited for comparison at a meeting of the Anthropological Society of Paris.

The announcement of the Neander Valley find came at a critical point in the history of science, only two years before the publication of Charles Darwin's *The Origin of Species* in 1859. The abuse heaped on Neanderthal man becomes understandable when viewed as part of the great philosophical debate on evolution. The idea of a changing prehistoric past, not to mention a primitive form of man, was a notion held by a tiny minority of scientists. The outlook of most educated people had scarcely altered from the Middle Ages, even though Kepler and Galileo had destroyed the old earth-centered universe. Belief in the divine act of creation, a fixed point in time, stood unshaken, although there was disagreement on the precision with which this date could be calculated. Many accepted the famous estimate of 4004 B.C. by the Archbishop of Armagh, based on the ages of Adam's descendants given in the Old Testament. This had later been refined by Dr. Lightfoot of Cambridge University, who considered that "man was created by the Trinity on October 23rd 4004, at nine o'clock in the morning."[2]

Within this framework of a literal reading of Genesis, scientists struggled to accommodate the record of geology. The fossil remains of creatures now unknown and extinct must be due to great "catastrophes" that, like the biblical Flood, had periodically deluged and destroyed whole populations of animals. Only cataclysmic disturbances of the earth could explain why primitive-looking stone tools were occasionally found mixed up with the bones of animals that had perished in the Flood. Yet as these discoveries became more frequent and better observed, a small number of researchers found it difficult to ignore the possibility of human existence during the age when elephants and rhinoceroses had ranged across Europe.

A key figure was Charles Lyell, whose classic, *Principles of Geology,* had been published in three volumes in 1830–33. In this work, Lyell marshaled the arguments that challenged the biblical timescale of creation and suggested that geological layers must have been deposited in an orderly sequence over an immense period. The ceaseless forces of seas and rivers, the intermittent activities of volcanoes and earth movements, could all account for the record of rock formations, without recourse to supernatural catastrophes.

Despite the influence of Lyell's work in scientific circles, particularly on the young Darwin, there were still many who preferred to disregard the evidence. For if the geological layers had been formed in an orderly pattern in the past, then there was an even more uncomfortable conclusion to be faced: rudimentary tools found side-by-side with the bones of extinct animals were indeed contemporary, and man himself had existed in remote times. Well-documented discoveries of hand-axes in France and Britain came to light only a year or two after the first accounts of the bones from the Neander cave. The combined effect of these finds stirred Lyell into action. Could the skeleton belong to one of the ancient toolmakers, and was it perhaps a primitive, apelike form for which Darwin's newly published evolutionary theories had prepared a place? Since Lyell was engaged in writing *The Geological Evidences of the Antiquity of Man,* it was essential for him to study the problem first-hand.

In 1860, Lyell visited Fuhlrott, who guided him around the Neander cave and presented him with a plaster cast of the skull cap. The bones, Lyell concluded cautiously, had the general appearance of fossil remains. His opinion was confirmed in the first full study of the skull, which was done by Thomas Huxley, to whom Lyell had given the cast on his return to England. Huxley's careful examination was published as part of Lyell's new book, and it established, among other key features, the large volume of the skull. This convinced Huxley that the Neanderthal remains were more closely related to modern man than to the apes. They did not represent a halfway stage in human evolution, but instead a prehistoric man with marked primitive features.

By the time *The Geological Evidences of the Antiquity of Man* was published in 1863, reactions to the original discovery had encompassed a remarkable range of views. On the one hand, Virchow and his followers, with their traditional outlook, poured scorn on the theory of evolution and on the Neander bones, which were nothing more than those of a modern freak. By contrast, Huxley's work claimed a comparatively late, fully human place for Neanderthal man in the line of evolution. For fifty years, the verdicts on Neanderthal fluctuated wildly between these two extremes.

Eventually, sporadic finds toward the end of the nineteenth century dispelled any lingering prejudices against accepting Neanderthal as a genuinely ancient form of man. The most notable discoveries included two skeletons at Spy, Belgium, found in 1886, and the fragments of at least thirteen at Krapina, near Zagreb, Yugoslavia, between 1899 and 1905. By this time, the theory of evolution had finally met general acceptance throughout Europe, and opinion tended to place Neanderthal man on the direct line of evolution that led to *Homo sapiens sapiens*.

From being an arthritic outcast, Neanderthal man had at last emerged with a claim to human recognition. Just what kind of humanity he represented, and whether or not modern Europeans

The discovery of the first of the burials at La Ferrassie in the Dordogne in September 1909. Note the straw boater, pick-axe and luncheon basket.

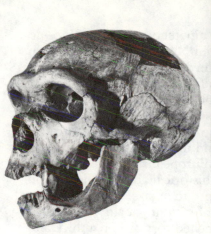

The skull from La Chapelle-Aux-Saints, and *(right)* a view of the tiny cave entrance in which the skeleton was found in 1908.

would have felt comfortable in his presence, was quite another matter, for in 1908 a discovery was made that opened up the most bizarre chapter in the eventful Neanderthal discovery. This time the setting was in the Dordogne region of southwest France, close to the village of La Chapelle-Aux-Saints. The interpretation that was made of the discovery by Professor Marcellin Boule of the Natural History Museum in Paris deeply influenced scientific opinions and struck a chord in the imagination of the public that was to last for decades.

The discovery in the tiny cave near La Chapelle-Aux-Saints consisted of the skeleton of a man of about fifty, laid to rest in a shallow, artificially dug grave. The man had been buried on his back, head facing the west, with the right arm bent and the legs drawn up toward the body. Around him lay numerous fragments of quartz, flint, ocher, and animal bones, but since the soil around the grave pit also was littered with such objects, it is difficult to know if they were deposited intentionally as offerings at the old man's burial. There were hearths inside the cave and many animal bones, including those of reindeer, bison, horse, and ibex, which must represent the remains of a number of meals. Today, the cave is an unpleasant, musty cavity; its ceiling is too low to allow one to stand upright, and this convinced the original excavators that it was "not a habitation place but a tomb, where people would have come to make many funeral feasts."[3]

The significance of the old man of La Chapelle-Aux-Saints preoccupied Boule for many years, and he published several detailed descriptions of the finds. In his attitudes toward the Neanderthal problem, Boule seems to have been influenced by conservative ideas, perhaps even by the catastrophe theories that had lingered among French scientists during the late nineteenth century. Since no immediate links could be found between Neanderthal and modern types, Boule envisaged that the Neanderthals had been forced into extinction—wiped out, like the fossils of remoter ages, by an invading "flood" of modern men—and had played no part in the progress of human physical and cultural development.

To support his point of view, he emphasized as many features of the Neanderthal bones that he could find that seemed to distinguish them from the skeletons of modern Europeans. In doing so, he produced a monstrous caricature of prehistoric man that quickly won a popular place in the history books. Despite its size, the brain of the Neanderthal was, Boule suggested, inferior in quality to that of modern man. This empty head was hunched forward on a thick and stumpy neck, supported by a spinal column so curved that a fully upright posture was impossible. The unfortunate Neanderthals could not extend their legs properly because of the limitations of their knee joints; because of the pronounced angle of their great toes, they were forced to amble like apes on the outer edges of their feet.

The remarkable fact is that every one of these physical traits has been systematically disproved by unbiased researchers during the course of the past sixty years; yet the popular image of the slow-witted hunchback has stuck fast.

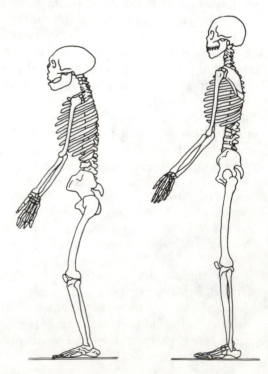

Boule's reconstruction of the skeleton from La Chapelle-Aux-Saints, contrasted to *(right)* a modern skeleton.

The Great Debate

After the remains of about 200 individuals have been discovered throughout Europe, North Africa, and the Near East, Neanderthal man emerges in a very different shape in the hands of modern anthropologists. Within this immense area, the characteristic features of the skeletons vary so much that it is doubtful that one descriptive term like "Neanderthal" has any real meaning. No specimen is so outlandish or subhuman in its physical makeup that it could not fall within the range of modern man. Indeed, the scientists who reexamined the La Chapelle skeleton in 1957 concluded their report by proposing that if a Neanderthal man

> could be reincarnated and placed in a New York subway—provided that he were bathed, shaved and dressed in modern clothing—it is doubtful whether he would attract any more attention than some of its other denizens.[4]

No evidence exists to show that they were not capable of a fully upright posture, and the 1957 investigators were able to prove that Boule had been misled partly by the arthritic condition of the old man of La Chapelle. "In reality," wrote one imaginative Russian anatomist:

> the Neanderthaloid stood very firmly on his own two legs, was mobile and nimble, he reacted with lightning rapidity and displayed a great elasticity in his coarse limbs.[5]

While the monster image of the past can be dismissed with confidence, the special character of certain finds, among them the earliest and best-known discoveries, does invite attention. These were undoubtedly stocky and muscular individuals. Their broad, rugged faces were characterized by a receding chin and forehead and jutting eyebrow ridges. The *average* measurements of these features fall well beyond the *average* of any human population that exists today.

Nevertheless, modern discoveries have made it clear that this rather striking physical type was restricted not only to western Europe but also to a limited period of time. During the mild period before the beginning of the Ice Age, the few bone remains available suggest far less physical variety than seems to have developed later on. Indeed, some of these early specimens are of such a slight and unpronounced character that they appear close to the modern average. Then, as the glaciers encroached on Europe again, from about 70,000 B.C. onward, there seems to have been a physical response to the changing conditions: a steady formation of distinct characteristics of a regional or racial kind. The western Neanderthals, such as those from Neander, La Chapelle, and Spy, were one such population, but they were rather exceptional. Their contemporaries elsewhere in Europe belonged to many diverse physical strains, most of them of a less rugged and extreme build.

Exactly what favored these divergences between populations is hard to say. It is often suggested that the "classic" Neanderthals of western Europe were heavily built to withstand particularly adverse temperatures and that perhaps the relative isolation of particular regions may have encouraged the gradual emergence of their distinctive traits. But the origin of the European peoples was not related simply to important changes in their surroundings. They were not passive backwoodsmen at the mercy of every fluctuation in temperature and vegetation. Instead, research shows that they participated in a remarkable episode of expansion and colonization: a pioneering role that is certainly at odds with their conventional image.

Vast regions of Europe lay empty of human settlement 100,000 years ago. At this time, in the midst of the mild interglacial conditions, formidable expanses of pine forests in most of central and eastern Europe undoubtedly barred the way to effective hunting. The dense cover of conifer forest could support meager quantities of game in comparison with the grassy plains farther south or even the tundra far to the north. When the first bitter phases of the Ice Age drew in, the forest frontier gradually receded, and the newly opened grasslands offered the people of the Neanderthal era new hunting prospects in virgin territory.

The scale and extent of the population movement that followed is difficult to grasp. It was certainly driven from a number of different directions both in the west and the east and must have taken place over the course of several thousand years. Most of Europe east of the Rhine and north of the Alps and the Caucasus was effectively occupied for the first time. Between the Alps and the Urals nearly 300 settlements are known from the early Ice Age, and it has been estimated that this is seventy times the number of sites recorded from all the previous periods put together. Yet this expansion of population did not stop even at the Urals but extended into the enormous region of Kazakhstan to the south and onward toward Siberia. Settlement seems to have halted, at least temporarily, on the fringes of the Siberian plateau, which was probably thickly forested.

In penetrating this far, the people of the early Ice Age must have been able to adapt to a remarkable range of living conditions. Their capacity for survival is demonstrated by a few scattered sites found recently just below the Arctic Circle on the western flanks of the Urals. These may represent only a brief episode of occupation during a comparatively temperate phase of the Ice Age, but the resourcefulness of the prehistoric hunters in withstanding the arduous winters in such regions can scarcely be doubted.

Adequate fur and hide clothing must have been essential to survival on the windswept plains of the north, although items of worked bone, such as buttons and bodkins that might indicate carefully fitted garments, are not in evidence. It is equally certain that efficiently heated artificial dwellings were vital in many regions of eastern Europe. Yet so far, convincing evidence of an early Ice Age hut has been recovered from only one site in southwest Russia, dating to about 42,000 B.C. The structure was uncovered at Molodova, near the banks of the Dniester in the Ukraine, and enclosed a living space measuring about eight by five meters. The basic framework of this hut was probably of wood covered with skins, which were weighted down at the edge by over one hundred mammoth tusks, skulls, and bones. Inside the hut, diggers uncovered fifteen hearths, many animal bones, and thousands of flint flakes, which indicate a long period of use. Traces of another structure at Molodova also were located a short distance away along the Dniester.

As interesting as these finds from the Ukraine are, they provide a pitifully inadequate basis for understanding how the vast regions of eastern Europe were opened up, what types of communities were involved, or how dense the settlement eventually became.

To establish a picture of the kind of society to which Neanderthal people belonged, specialists have concentrated on the evidence of their flint tools, found in abundance wherever caves, rock shelters, or open-air camps were occupied. A few Neanderthal bodies were buried alongside a distinctive range of tools that take their name from the site where they were first noted, Le Moustier in the Dordogne. The so-called Mousterian tradition of flint-working is found

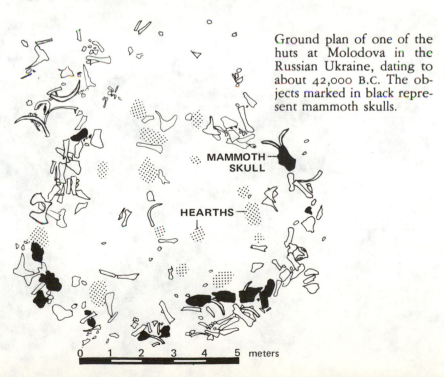

Ground plan of one of the huts at Molodova in the Russian Ukraine, dating to about 42,000 B.C. The objects marked in black represent mammoth skulls.

MAMMOTH SKULL

HEARTHS

0 1 2 3 4 5 meters

in much the same broad areas as the remains of Neanderthal man, from western Europe and northern Africa as far east as central Asia.

The expert on flint tools is confronted not only with an enormous geographical range but also with a great period of time. Tools of a recognizably Mousterian form appeared in Africa and Europe by at least 150,000 B.C. or earlier, although the tradition was only widespread by a very much later date. Indeed, the flowering of the Mousterian industries seems to have been connected with the great expansions of population that happened toward the beginning of the Ice Age, from about 70,000 B.C. onwards. At this time—just as the biological form of the prehistoric Europeans diverged into a wide variety of local populations and races—a multitude of particular variations on the Mousterian theme characterized the tool kits of certain regions.

Obviously, the study of these traditions is a complicated matter for the academic and the specialist; why, then, is the subject such an important one? It may be wondered how the technical evidence of flint tools can ever give us an insight into these scarcely imaginable lengths of time. In fact, the stone tool debate vitally affects our entire attitude toward the Neanderthals.

To begin with, experts on Mousterian tools recognize more than sixty basic, intricately finished types. Many of these tools called for delicate and sensitive retouching after the basic form had been struck from the flint. It takes weeks of practice for a modern enthusiast to

A selection of Mousterian tools belonging to several different industries.

reproduce the ancient techniques to the same standard of finish. It is obvious that the fingers of Neanderthal people were dexterous and that the complexity of their practical tasks required a refined and diverse tool kit. Indeed, the demand for raw material drove at least one Mousterian community to batter lumps of flint from the natural seams visible at a limestone outcrop at Löwenburg in the Swiss Jura. This site, which dates to about 40,000 B.C., is probably the oldest open-cast mine in Europe, and reflects something of the importance attached to stone-working by the Mousterians.

During the past twenty years, our knowledge of these tool traditions has deepened, largely as a result of the efforts of one outstanding figure among French prehistorians, Professor François Bordes. He has reached some remarkable conclusions about Mousterian society, based, in the first place, on a long familiarity with the tools themselves. Bordes has been able to manufacture prehistoric tools expertly since he was a boy and has excavated several major Mousterian sites in the Dordogne region with greater care and precision than had ever been attempted before. Bordes' fundamental achievement, however, was to devise the basic categories into which the tools are now universally recognized and divided in western Europe. Once it became possible to describe the contents of particular tool kits both accurately and systematically, it was easy to compare one set of tools with another.

This led Bordes to the remarkable conclusion that the Mousterian layers at his sites contained evidence not of one developing culture but of four major groups existing side by side. The indications came from a simple analysis of the number of tool types present in each layer, which does not follow a random pattern. Instead, the frequency of each variety of tool in any one layer follows definite rules, allowing each layer to be assigned to one of the four Mousterian cultures. What this means in practice, according to Bordes, is that the people who made tools in the "Typical Mousterian" style—for example, at Combe-Grenal—apparently followed customs different from contemporary craftsmen manufacturing tools in the "Denticulate" style at Pech de l'Azé, only 10 kilometers away across the Dordogne. Moreover, the "Typical Mousterians" of Combe-Grenal concentrated on hunting reindeer during the same period when the "Denticulates" of Pech de l'Azé seem to have preferred horses.

It is tempting to imagine that these Ice Age communities were isolated and cut off from frequent contact with one another, thus encouraging the growth of local traditions. But the implications of Bordes' work are more far reaching than this: it seems that the four main traditions coexisted over wide areas of southern France and, some have maintained, throughout much of Europe and the Near East as well. How could prehistoric people living at one site practice

American graduate students digging at Pech de l'Azé in the Dordogne region.

their toolmaking in a manner identical to others, perhaps dozens or hundreds of kilometers away, and yet century after century remain oblivious to the quite different customs of their closest neighbors? To explain this, Bordes and his followers claim that the different tool groups correspond to separate ethnic or racial groups. So, for instance, the makers of the "Typical Mousterian" did not mix with the "Denticulates," nor did the "Acheulian type" mingle with the "Charentians." Instead, the distinctive identity of each of these separate groups was expressed in their toolmaking habits wherever they settled in western Europe.

Looking imaginatively at the complex statistical data presented by Bordes, we can appreciate that his evidence has elevated the Neanderthals to a status far removed from the "degenerate" race of the earlier studies. Bordes' Neanderthal is a self-conscious creature with a clear notion of a social identity beyond himself. While his crafts and beliefs have spread across Europe, in the Dordogne he pursues a settled and comparatively secure existence, with a distinct tribal consciousness and perhaps even loosely defined territories.

It is not surprising that Bordes' conclusions, which imply so much social complexity in Mousterian times, have been the subject of intense controversy. The major attack was mounted by a husband-and-wife team, the Americans Lewis and Sally Binford, who favor a simpler view of Mousterian society. Their work implies that the tool kits may after all belong to one developing Mousterian culture. Could the diversity apparent among the industries be due not to ethnic or cultural customs but to other factors? Could it be related, for example, to different activities that arose because of the varied practical needs of the hunting community from one season to the next? Suppose that at one type of site or at one particular time of year, butchering was a more common activity than scraping hides or woodworking; would this not result in the kind of differences in discarded tools observed by Bordes? With an ingenious and original application of statistics, the Binfords tried to identify repeating combinations of tools common to all four main groups that might correspond to the effects of particular activities or environments.

The upshot of the Binfords' work was to advance a less sophisticated picture of the Neanderthals than was implied by Bordes. Their Neanderthal has no concept of an ethnic identity beyond himself, and his craftsmanship is not the product of self-conscious custom. Instead, its character is determined largely by practical demands and the Neanderthal's responses to them, whether these are

Professor François Bordes relaxes at Pech de l'Azé while (in the foreground and at the right) his colleague Jacques Pelegrin trims the edge of a hand-axe with an antler hammer.

related to a certain season or to a particular setting. Thus were polarized two quite different explanations of the variety present in Mousterian tools, involving two contrasting conceptions of Neanderthal man.

This debate, which began modestly in the mid-1960s, soon developed into a classic academic confrontation. In 1965, the Binfords met Bordes in France, and Lewis Binford recorded his own impressions of the event in glamourized terms reminiscent of a prize fight. As Sally Binford observed, they began "like two male dogs circling and sniffing trying to find out if the other was really an enemy." Lewis Binford continues his account:

> We must have argued for more than an hour, our voices getting higher and higher, standing up, sitting down, pacing back and forth, leaning over the charts, big clouds of smoke issuing from his pipe, equally big clouds pouring from my cigarettes. . . . All of a sudden Bordes jumped up and came around face to face. I stood up almost automatically. He put his hand on my shoulder, looked me directly in the eyes, and said, "Binford, you are a heavyweight; so am I." I put my hand on Francois' shoulder; he turned: "Let's go and drink some good wine."6

The great debate was not at an end, however, for it continued in numerous less personal exchanges of academic papers. Despite the value and originality of Lewis Binford's early work, his later writing grew increasingly incomprehensible and full of jargon, which tended to obscure important flaws in his argument. Indeed, the heat generated by the confrontation seems quite out of proportion to the tiny number of properly excavated Mousterian sites that were under consideration.

Into this furor a new point of view was offered in a characteristically British spirit of compromise. Paul Mellars, who also had studied the Mousterian tools first-hand, voiced certain crucial objections to the Binfordian argument but also supported it to a limited extent. Some elements in the tool kits, he agreed, *were* probably responses to particular practical jobs. However, the main source of the patterning in the tool kits was a simple chronological one: the main Mousterian groups had succeeded one another over the immense time span of the middle Paleolithic. Mellars did not attempt to deny the possible existence of a sense of ethnic identity among the Neanderthals, but his essential argument was that *some* of the Mousterian variations represented one culture that had gradually changed its character over time. This suggestion clearly represents a less extreme view of Neanderthal man than was advanced by either Bordes or the Binfords.

After more than a decade of intense disagreement, where does the verdict on the Mousterian problem stand today? As we have seen, Bordes' point of view involves the division of Mousterian France into several distinct tribal groups. This seems plausible enough along the Mediterranean coast, where particular clusters of caves and open-air sites, all dominated by one variety of Mousterian, occupy individual blocks of land adjoining the sea. These do have the appearance of distinct tribal territories. However, in southwest France, the situation is quite different, for there several types of Mousterian occur at a single site. If we agree with Bordes, it is necessary to imagine tribes or races perpetually jostling one another in this small region, all pursuing the same way of life century after century, yet never influencing the traditions of their close neighbors. This seems highly improbable.

It is difficult to test whether ethnic or racial affiliations existed among the Neanderthals. Indeed, a few of Bordes' statements have the unspecific quality of a Middle Paleolithic novel:

> If a woman from the Quina-type Mousterian was carried off by an Acheulian-tradition Mousterian man, she may perhaps have continued to make her tribal type of thick scraper . . . but after her death probably no one went on making them. And finally, it must always be remembered that the Palaeolithic world was an empty world. . . . A man must often have lived and died without meeting anyone of another culture, although he knew "that there are men living beyond the river who make hand-axes."[7]

This type of reconstruction may appeal to our imaginations, but it is on the level of prehistoric fantasy, not science.

On the other hand, there is equally little practical evidence to support the Binfords' contention that the Mousterian tool patterns were the result of particular activities or environments. It is true that at least one type of tool kit, the "Denticulate," which is dominated by sawtooth flints that would surely have been useful for woodworking, appears in all sorts of places at all sorts of times, and so may indeed reflect practical requirements more than anything else. But the other industries *do* fall into varying degrees of chronological and geographical order; for instance, the "Acheulian-type" Mousterian is virtually absent in southeast France but appears at almost every site in the Dordogne. Why should a basic activity have been important in one region and not in another?

The answers to the Mousterian problem are clearly not simple, and they may well lie outside the traditional territory of French and American scholars. A new study of the Mousterian cultures in central and eastern Europe shows how much potential such regions have for the understanding of these issues. It is shown that a minimum of twenty separate toolmaking traditions flourished during the period of the early Ice Age, from about 70,000 to 40,000 B.C. The author of the report, Míklos Gábori, then provides detailed evidence to support his conviction that no correspondence can be found between the varieties of tools and such obvious factors as climate, vegetation, or hunting practices.

Probably the fairest judgment, in the absence of so much essential information, is that it would be unwise to exclude the possibility of tribal custom and craftsmanship from the world of the Neanderthals. At the same time, it is essential to pursue the tests that might reveal the effects of particular activities or of developments through time.

The Cave of the Grasshoppers

One of the best examples of how this problem can be approached was the study of a site in France, published in 1972. The report on the site of L'Hortus is a fascinating and intricate portrait of the life of a group of Neanderthal hunters in the Languedoc region of southern France over the course of 20,000 years. The book also is a monument to the rapid advances made in the precise reconstruction of Ice Age environments and the human responses to them, consisting of a collaboration of no less than thirty-six specialists led by Henry de Lumley of the University of Provence. A prodigious amount of detail about the prehistoric landscapes and activities is extracted from such unlikely sources as the types of grasshoppers that cooled themselves in summer among the chimneys of the Hortus cave and the species of trees that are present in the charcoal from numerous hearths around which the Neanderthal occupants gathered for warmth.

The hunters were perched high on a precarious rock ledge, where they looked down over a stony slope leading to the valley bottom over 200 meters below, while one hundred meters of limestone cliff face soared above them. This ledge was cut off from the interior of the cave by an abrupt natural fault in the limestone over two meters across. The Neanderthals often climbed into this narrow crevasse to light fires or to store bodies of animals which they had killed.

These animals were principally the ibex, which must have abounded on the rocky terraces around the cave, but other species, such as deer, reindeer, boar, and horse, were hunted occasionally in the forest bottom. The Neanderthals also stalked young carnivores, such as panthers and wolves, almost certainly to obtain their furs. Their clothing could have been made colorful by the plumes of choughs, partridges, and pigeons; at any rate, breakages in the bones

The cleft in the limestone near the entrance of L'Hortus into which the Neanderthals often crawled to light fires or to store dismembered ibexes and other animals.

of these birds correspond to the damage that would be caused by plucking. Fully grown lynxes and smaller game, such as foxes, rabbits, and birds, were likely to have been caught in traps. Despite the versatility of their hunting techniques, the Neanderthals of L'Hortus appear to have had no interest in fishing or snail collecting. In their lofty home high above the stony gorge, their water supply must have been a problem, and presumably, it was brought up regularly from the valley in some kind of container. Firewood also was collected down below, and for this purpose, nonresinous trees, which would burn efficiently and cleanly, were always selected.

The importance of the Hortus study lies not merely in these intimate details of everyday existence but in the comparison of dozens of separate living floors that succeeded one another throughout 20,000 years of occupation. These comparisons have allowed changes in types of climate, vegetation, animals, and human hunting strategies to be traced and interrelated in absorbing detail. There is strong evidence to show that the Neanderthal settlement in the cave was always intermittent and temporary, taking up a few months of the year. As the centuries passed and the climate changed, so, too, did the nature of the Neanderthals' activities and the seasons at which they visited the cave.

To begin with, this occurred between January and March, when it seems that the Neanderthals occasionally gathered under the western porch of L'Hortus to dismember the fully grown male ibexes that they had caught in the hunt. At a later period, their stops became more frequent and prolonged, lasting from January into April or May. To judge from the pattern of bone remains, they had ceased using the cave mainly as a butchering place and instead engaged in a broad range of domestic activities.

The final major phase of occupation coincided with a swift drying of the Mediterranean climate and the shrinking of forest cover in the valley below. In these very arid conditions, there is nevertheless pollen from aquatic plants preserved in the cave, and this is probably evidence of vegetable foods brought up from the valley by the Neanderthals. They now came to L'Hortus in the summer (most of the grasshopper population deserted the cave). They still hunted ibexes, but a wider range of quarry was pursued. The occupation of L'Hortus seems to have reached its climax: distinct pathways trodden through the rubbish in most parts of the cave can be detected, and numerous animal jaws were crushed by the pressure of human feet.

So the character of the temporary visits to L'Hortus changed between 55,000 and 35,000 B.C. From an occasional butchery place, it became the site of more frequent and general activities. But the flint tools struck by the Neanderthals remained almost unchanged throughout the whole period, and all belonged to the "Typical

Mousterian" variety. There is no relationship detectable between the changes in climate, vegetation, or human activities and the relatively constant flint-working habits of the cave's occupants. This is a conclusion of vital importance, a compelling reason for thinking that fixed notions of craftsmanship were passed down from generation to generation in the Mediterranean area.

The Awakening Spirit

In addition to the regular ibex feasts, there were, it seems, some rather grisly additions to the menu enjoyed by the Neanderthals at L'Hortus. From seven separate living floors, human fragments representing a minimum of twenty people have been recovered and examined. These all display somewhat slighter features than are typical among the big-boned Neanderthals of the west, but their basic form is similar.

The bone fragments convey more than just anatomical information, however. The range of ages represented is not what we would expect, for there is a high proportion of remains belonging to people over fifty and few from under fifteen. Furthermore, most of the bones are skull and jaw fragments. For some reason, the other parts of the skeletons have disappeared, despite the fact that in the same layers all the joints of animals are well-preserved. The human fragments are in fact mixed with the animal bones, suggesting cannibalism. Two human long bones show intentional breakages that are identical to those on animal bones, undoubtedly made in order to extract the marrow. Considering the wide range of animals that was hunted at L'Hortus, it is unlikely that these activities were pursued for nutritional needs. Indeed, a strong hint of ritual came from one layer, where a mass of debris, including various human bones, three ibex skulls, and fragments of four lynxes, appears to have been covered over by a wolf skin, with the skull, jaw, and lower limbs of the wolf still attached to the fur. Was this the remains of a bizarre hunting ritual?

Here is the most convincing evidence so far advanced that Neanderthals dabbled in some form of cannibalism. Similar indications (often poorly observed or recorded by excavators many years ago) are known from a number of other Mousterian sites. At Teshik Tash in Uzbekistan, for example, the jumbled skeleton of a nine-year-old child was found to bear cutting marks where the flesh had been removed from the body. The remains were mixed with ibex bones, and six pairs of ibex horns had been placed in a ring around the child's skull.

An equally dramatic discovery occurred at the Guattari cavern on

Monte Circeo, a little over one hundred kilometers south of Rome, which had been sealed off from the outside world for more than 50,000 years. In 1939, A. C. Blanc penetrated the cavern and found himself walking on an intact Mousterian floor that had lain undisturbed since the middle of the last Ice Age. In one chamber, he found an isolated skull surrounded by a circle of stones. The skull belonged to an individual who had received one or more fatal blows on the head. The teeth were smashed out at the roots, and the base of the skull had been cut open, presumably to extract the brain. Since no trace of the neckbones or other parts of the skeleton was found, the head of the victim had probably been removed outside the cave after the body had already decomposed. An isolated fragment of another human skull, this time a jaw, was discovered elsewhere in the cave. A recent study of the hundreds of animal bones that litter the Mousterian soil has drawn attention to a curious selection of certain body parts to the exclusion of others and to the fact that many bones are burned, although no traces of a hearth exist anywhere in the cave. This evidence, together with the uncomfortably low ceiling, suggests that the cave was not an everyday living place but instead was frequented for obscure and perhaps gruesome rites.

The "bestial" side of Neanderthal man may strike us as distasteful, but at least it is plain that the beliefs associated with death were complex. At Shanidar cave in Iraq, nine Neanderthals were uncovered during the 1960s, several of whom had been struck down by boulders tumbling from the roof of the cave. The fate of several of the victims seems to have aroused a surprisingly delicate response among the living occupants of the cave. Ashes and food remains discovered among the stones placed over the Shanidar burials may indicate that funeral feasts were held in honor of the dead. The most remarkable discovery was that one burial had occurred at the end of May or early July, as determined from the pollen traces of eight different types of brightly colored wild flowers that accompanied the burial. These flowers were probably woven into the branches of a shrub to form a kind of screen placed with the dead body by Neanderthals who were perhaps more sensitive than their conventional image allows.

It is difficult to decide whether evidence of this kind testifies to simple reverence for the departed or reveals anything as complicated as a belief in an afterlife. One problem is that the majority of Neanderthal burials was discovered in the early years of this century, so records of the finds are anything but satisfactory. To cite one example, an unscrupulous Swiss digger, Otto Hauser, who came across the body of a youthful Neanderthal at Le Moustier in 1908, is recorded as periodically covering up the skeleton with earth and "rediscovering" it for the benefit of visiting dignitaries. Even in the few cases where skeletons have been brought to light under careful supervision, it is difficult to be sure whether fragments of animal bones, flint, or ocher were deliberately placed with the body or simply trodden in from the surrounding surface.

In some cases, the flexed position of the skeleton, with one hand often drawn up to the face, suggests an attitude of sleep or repose, which we can imagine was a ritual preparation for a reawakening in another world. On the other hand, some bodies appear to have been so constricted in their position that they may have been tied up with a cord or thong, possibly indicating a fear that the departed might walk beyond the grave.

The spirits of dead humans may not have been the only supernatural forces that Neanderthal people appear to have reverenced or feared. Long before the start of the Ice Age, hunters in the Alpine regions of Europe and elsewhere had to compete for their possession of winter homes in caves with the huge and powerful cave bear. Their respect for such a formidable adversary is apparent from several reported cases of bear skulls and long bones placed "on display" in clefts and ledges inside caves, sometimes covered by stone slabs or by layers of ashes.

These early observations were dramatically confirmed in the early 1960s at the cave of Regourdou in southwest France. The excavations first revealed a human burial, laid on a bed of stones next to flint tools and a bear bone. The Neanderthal body was covered over by stone blocks and a heap of sand, pebbles, and ashes, which also contained bear bones. Below this interment, in the earlier layers of occupation, was a series of remarkable structures, some of them associated with bear remains. Of these, the most extraordinary was a stone coffin, roofed by a great slab weighing around 850 kilograms. Under this lid was found the skeleton of a bear, complete but not in any natural anatomical order. It seems likely that bears (whose bones have a striking resemblance to human ones) were the object of superstition or veneration, as they were among many recent hunting peoples of the north.

Thus, Neanderthal man emerges from the record of prehistory with a recognizably human range of needs, anxieties, and beliefs. The preoccupation with skulls, the burials with food and plant "offerings," the likely existence of tribal forms of toolmaking, all indicate a new level of self-consciousness in Neanderthal people. It is natural that we should look in the Mousterian period for the first expressions of self-awareness in terms of art and ornament.

Not a trace of cave painting or engraving, neither a solitary splash of color nor a single hesitant scratch, can positively be identified as the work of Neanderthal people. Yet, curiously enough, the same raw materials used by the succeeding cave artists are found quite frequently in Mousterian layers, with few clues of the purposes that they could have served. For example, over one hundred fragments of the black mineral substance manganese dioxide were discovered at Pech de l'Azé, most of them worn flat on one side or polished into a rounded point as if they had been used as crayons to mark some soft surface. Other coloring materials, such as red ocher, are less common. Neanderthal people probably expended their creativity on materials that have not survived, such as wood and hides or perhaps on each others' bodies. It seems that the Neanderthals had neither the time nor the taste to spend on making even the simplest ornaments, such as necklaces and pendants. Only two simple pendants are recorded from the Mousterian period, both from La Quina in the Charente region, and since the site was excavated a long time ago, it is conceivable that they are from a disturbed layer and really belong to a later era.

However, there *are* unmistakable signs that the Neanderthals felt artistic impulses of a rudimentary kind. For example, they continued to manufacture the mysterious stone spheres that had been made by

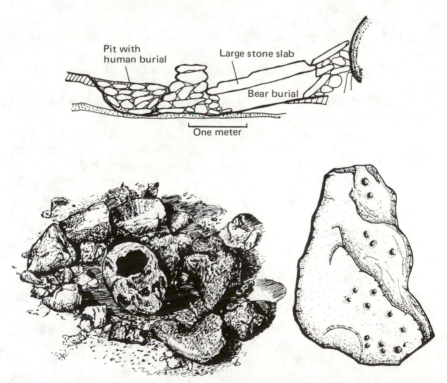

Evidence for Mousterian rites and beliefs. At the top, a cross-section of the extraordinary bear burial at Regourdou in the Dordogne. Left, the Monte Circeo skull as it was found lying on the floor of the cave. Right, the cup-marked slab from La Ferrassie, the oldest sculptured stone known.

A bone with engraved markings from Mousterian layers at the site of La Quina in the Charente district of western France.

their ancestors in the African continent for over 1 million years. Some are only two centimeters in diameter, scarcely large enough to be effective missiles, and in the Mousterian deposits, these balls are sometimes found in considerable quantities. No less than seventy-six were recovered from digging at La Quina, and from this same site came a great flat white disc of flint, perfectly worked to a diameter of over twenty centimeters, which can have had no conceivable practical function. These discoveries suggest that Neanderthal people attached some importance to the circular form and that, as their forebears in other ice ages must have done, they received pleasure from the texture and symmetry of certain objects that they prepared.

A remarkable find from Hungary, excavated at the Mousterian site of Tata and dated to about 50,000 B.C., very clearly illustrates these preoccupations. The craftsman at Tata split apart the corrugated, ridged surface of a mammoth molar in order to obtain a roughly oval section of the ivory. This segment of tooth was cut, carved, and handled until it reached a brilliantly polished state. One researcher, who recently examined the piece and found that it had at some stage been stained with red ocher, refers to it as "the most sensuous of all Palaeolithic mobiliary artifacts both to the touch and eye."[8]

Less spectacular carved objects—usually fragments of bone engraved with crude zigzags—are recorded from a number of Mousterian sites. Only one piece of carved rock, however, has so far come to light. It is an uninspiring item in itself, but it occupies a place of honor as the earliest example of monumental art known in Europe. It was discovered in an inconspicuous rock shelter in the Dordogne region known as La Ferrassie, not far from the little town of Le Bugue. Here, in 1909, the prehistorian Denis Peyrony came across the first human remains that belonged to a complex Neanderthal cemetery. In the course of the years that followed its discovery, one male, one female, and at least four children (including one that was apparently a fetus) were disinterred from a series of mounds and ditches covering the site. In 1973, the fragments of another child of about two years were recovered from the back of the shelter. Unfortunately, it cannot be determined whether the burials occurred over a long period or whether La Ferrassie was the last resting place of a single family.

The most remarkable find on the site came from a shallow pit, at the bottom of which lay the headless body of a child of five or six years, laid to rest in a tightly flexed position. Only about one meter away and higher up in the same pit was found the skull of the dead child, its jawbone missing, lying below a limestone slab. The underside of this slab was stained with red ocher, and its surface was pitted with eighteen shallow depressions or cups that were arranged in pairs and that were probably hollowed out with the help of a stone implement. We can only guess why the young child of La Ferrassie had been singled out for such an exceptional rite. Peyrony thought that the skull had been mutilated to allow extraction of the brain or so that it could serve as a ritual vessel.

The pockmarking of the slab at La Ferrassie may not seem worthy to be called art; yet in this simple action of grinding out a cup in stone must lie one of the rudimentary satisfactions of the creative urge. The cup appears over and over again as the basic element in the rock art created by primitive communities of farmers as well as hunters throughout the ages, for whom the symbol undoubtedly concealed countless different meanings. Yet Neanderthal man seems never to have advanced beyond this initial step toward self-expression. No example of representational art has yet been associated with him. The process by which these crude abstractions were transformed "abruptly" into the extraordinary vision of the cave artists is surrounded with mystery.

4: The Innovators

The Last of the Mousterians

The Mousterian horizon is very murky. And if those great eyes of theirs staring from under the arched brows could look into the future they would see nothing there, nothing but oblivion. They have had their day. The omens are against them now.

. . . The enemy on the horizon . . . they appear to be all painted red. It's red ochre surely? War paint, or I'm much mistaken. Tall fellows: a number of them stand as much as six foot four. They look fine in barbaric crowns set with the canine teeth of deer or studded about with bright shells. Their broad faces and long, narrow eyes, their high cheek-bones and eagle noses are quite handsome after too much of the gorilla-browed . . . Mousterians.

. . . And just when they and the Mousterians are coming into contact darkness descends. War whoops, yells in the night . . . we can only guess at what is happening in the turmoil everywhere . . . massacres, ending in the Mousterians' downfall . . . even their total extinction.[1]

We might be forgiven for thinking that the above extracts belong to a screenplay for *The Last of the Mousterians.* In reality, they are taken from a popular book called *Prehistoric Man* by Keith Henderson, published in 1927. The lurid scenario accurately reflects the state of expert opinion at that time on the fate of the Neanderthals and the rise of the cave artists. Scientists now recognize that the truth is more complicated and intriguing than was once supposed.

Until recently, many researchers were content to follow the evidence advanced by Boule and others, which indicated that the subhuman Neanderthals were pushed into extinction by a race of men invading from the east. These people were the first individuals to reach western Europe who were identical in all physical respects to modern man. The luckless Neanderthals were no match for the clean-limbed, upright, adaptable *Homo sapiens sapiens,* with his superior range of tools and weapons, his skill in hunting, and his developing powers of artistry and self-expression. The competition for game either forced Neanderthal people out of their favored hunting grounds into marginal territory where survival was precarious or they were deliberately wiped out in a full-scale racial confrontation. In either case, modern Europeans were definitely spared the uncomfortable possibility of having Neanderthal genes. The unconscious appeal to racial pride, which made this explanation popular for so long, was made quite explicit by Boule, who suggested that Neanderthal man

> represented a belated type existing side-by-side with the direct ancestors of *Homo sapiens;* its relation to the latter was similar to that which exists at the present day between the races we call inferior and the superior races. Perhaps one might go so far as to say that it was a degenerate species.[2]

Another prehistorian of Boule's generation, R. A. S. Macalister, provided a stark summary of the extreme viewpoint that prevailed during the first half of this century:

> Just as the early colonists in Tasmania used to organize *battues,* in which the unfortunate aborigines were the game, so the incoming Upper Palaeolithic people "shot at sight" whenever a Mousterian man made his appearance, until the ancient race was almost wiped out.[3]

So the origins of modern man came to be described in terms of the recent colonial past, with the cave artists cast in the role of empire-building aggressors. The persistence of this view in some popular books ignores fascinating recent evidence on the question of how and why the society of the cave artists (the "Upper Paleolithic") originated.

Some of the most interesting discoveries were made in the winding valley of the River Cure, about 200 kilometers from Paris. Near the village of Arcy-sur-Cure, the river has eroded nine caves in the limestone cliff, notably the Great Cavern with its spectacular stalactites, which were the object of curiosity to many distinguished sightseers ever since the visit of Louis XIII in the seventeenth century. Several of the less conspicuous caves have attracted equal interest in the eyes of archaeologists, for their entrances have concealed rich deposits from the Mousterian and later periods. In particular, the Cave of the Reindeer at Arcy was the site of a remarkable fifteen-year campaign of excavation by French prehistorian André Leroi-Gourhan, who discovered the cave in 1949. His deductions throw a vivid light on the problem of what happened to its Neanderthal inhabitants.

Leroi-Gourhan demonstrated the existence of some fourteen separate layers in the debris that had accumulated at the entrance porch of the cave, totaling more than six meters in thickness, almost half of which spanned the vital period of transition between the Mousterian and the Upper Paleolithic cultures. Had the invading hordes of modern men swept all before them, as the conventional picture suggested, one might have expected a break in the occupation of the Reindeer Cave to be evident, such as an interval of several decades or centuries when the site had been abandoned. In any event, even if the newcomers had swiftly taken over the cave mouth after its desertion by the Mousterians, some decisive change in the nature of the rubbish dropped at the site would surely be detectable. In point of fact, the meticulous excavation of the Reindeer Cave layers revealed no such dramatic change but instead indicated a steady development toward the new customs and techniques practiced by the modern men of the Upper Paleolithic. This was a crucial discovery.

Instead of revealing the eventual loss of nerve in the beleaguered survivors of a nearly extinct race, the Final Mousterian layers at Arcy seem to show a distinct advance in the capabilities of Neanderthal people. For the first time, fragments of bone and ivory were frequently worked for the production of spear points and other objects. While they continued to manufacture flint implements with the same basic techniques that their ancestors had practiced for thousands of years, the Final Mousterians began to develop a more delicate finish and a more refined range of shapes that foreshadow the tools of the succeeding Upper Paleolithic cultures.

Moreover, the Final Mousterians collected objects that aroused their interest, such as fossilized mollusk shells and oddly shaped lumps of iron pyrites. They were presumably picked up in the course of hunting forays and brought home to the cave as curiosities or perhaps for magical purposes. This may represent the first stirrings of an artistic urge among the prehistoric inhabitants; in any case, the objects do clearly demonstrate an awareness of natural forms and an inquisitiveness about peculiar fossils and minerals that was commonplace among their Upper Paleolithic successors.

The real interest of the Reindeer Cave, however, is the layers that followed immediately on top of the Final Mousterian. The inhabitants of the cave porch during this era (probably in the centuries after about 31,000 B.C.) left behind traces of huts constructed around a rigid framework of mammoth bones and tusks. The use of these materials probably reflects the scarcity of wood during this period of dry and cold conditions, when the only tree cover to interrupt the barren grasslands was that of sparse copses of pine, juniper, hazel, and alder, which were probably growing in the shelter of the Cure valley.

To protect themselves from their bitter environment, the cave inhabitants took to their bone huts, which must have been covered with skins and warmed by a central hearth. Some of the floors enclosed by the bone remains consist of well-trodden, ocher-stained

Unusual objects collected by the Mousterian occupants of the Hyena Cave at Arcy-sur-Cure in about 35,000 B.C. The lump of iron pyrites, the fossil gastropod and fossil coral were all found together.

earth. The lumps of ocher discovered in the hearths imply that the occupants of the huts almost certainly knew how to produce variations in the shades of color by heating up the raw material, which was subsequently ground into powder on flat "pallet" stones.

Further evidence of the developing artistic sense of the inhabitants is their new taste for personal ornaments, including simple pendants made of disks of bone, small fossils, and animal teeth with holes or grooves suitable for suspension. Other pieces of bone were marked with series of spaced lines or notches, which may represent the first-known examples of a tallying or counting system. Indeed, the Arcy bones are the earliest examples yet discovered in Europe of this custom of engraving regular notches on bone, which archaeologists refer to by the rather dubious term "hunting marks" *(marques de chasse)*. The practice of engraving these bones does imply the observation of some phenomenon over a period of time, but their exact purpose is a matter of controversy; these theories are discussed in detail in Chapter 12.

The hut foundations, the abundant quantities of ocher, the pendants, and the marked bones obviously reflect an accelerating social development among the prehistoric population at Arcy. The contrast to the previous Final Mousterian layers is heightened by the fact that, for the first time, the inhabitants of the Reindeer Cave bothered with regular "housework." They systematically tidied up the living floor on which the Mousterians had been content to sit, surrounded by the putrefying remains of their meals, and deposited their rubbish in orderly heaps outside the cave. But who were these hygienic successors to the Mousterians—Neanderthals or modern men? And what exactly had happened in the transition between the two stages at Arcy?

The people who built the huts and wore the pendants also produced a distinctive style of flint-working. One of its most characteristic pieces is a broad, tapered blade with one side worked into a

Animal bones discarded by the Mousterians in the Hyena Cave at Arcy.

curved blunted back. This is known as the Châtelperron knife, and it gives its name to the whole flint-working style, the Châtelperronian. It is this style that was once thought to be the hallmark of the invading *sapiens*. Here was the first appearance of the new, refined fashion of long-bladed tools, heralding the arrival of the Upper Paleolithic in France.

While some features of the Châtelperronian flints *do* hint at the tools of later hunters, a critical reexamination of those from such sites as Arcy show how much they also owe to the implements of the Mousterians, especially the Châtelperron knife itself. Everything suggested that the essentials of the Mousterian culture had survived, although crucial developments had occurred in their society and technology.

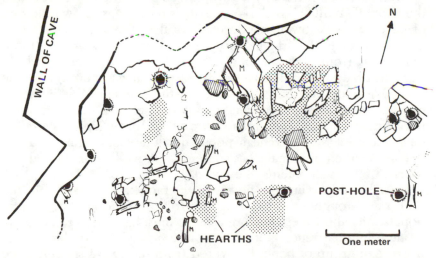

Plan of a hut built by the Châtelperronians at the Reindeer Cave, Arcy-sur-Cure. A ring of mammoth bones (M), river pebbles (striped), paving slabs and eleven post-holes surrounds the central living area inside the hut.

Yet the racial identity of the Châtelperronians remains a greater mystery than ever. An almost complete sapiens skeleton was found at Combe-Capelle in the Dordogne in 1909, supposedly in a Châtelperronian layer. Perhaps, then, the Châtelperronians were, after all, the "new" modern men who, if they were in any sense "invaders," nevertheless adopted some of the ancient traditions and habits of the Mousterians. Unfortunately, the discoverer of the Combe-Capelle skeleton was none other than the unscrupulous Otto Hauser, fresh from his devious unearthings of the Moustier Neanderthal a year previously, and so little reliance can be placed on his observations of the exact location of the skeleton in the layers at Combe-Capelle.

In contrast, experts who examined the Châtelperronian deposits of the Reindeer Cave at Arcy considered that a number of human teeth found there were closer in character to Neanderthal than to modern teeth, although dental indications alone are not conclusive. If the Châtelperronians were in fact Neanderthals, it is clear that they were well on the way toward developing an Upper Paleolithic way of life.

Inheritors or Exterminators?

A decisive or violent break with the Mousterian past—a systematic campaign of genocide, as some imagined fifty years ago—is unlikely to fit so complicated a picture. The character of the Châtelperronian flints is a persuasive argument in favor of some degree of continuity with the past, while any conclusions of any kind are much harder to draw from the fossil bone record.

Several recent studies by anthropologists reinforce the view that the ruggedly built "classic" Neanderthals of the west show little sign of developments toward modern physical features. An interesting case in point is the fragmentary remains of the twenty or more individuals from L'Hortus in southeast France who seem to have been the object of cannibal feasts at separate intervals throughout the 20,000-year occupation of the cave. In the course of this immense period, the bones show no trend of any kind toward the characteristics of modern man. The inhabitants of L'Hortus almost certainly belonged to a Mediterranean population of Neanderthals that remained in their regional "backwater," relatively isolated and untouched by developments elsewhere.

Outside western Europe, however, the range and variability of the skeletons throughout the Ice Age are striking. Among these diverse populations, there were individuals who must have been remarkably modern in appearance, notably in Palestine and North Africa. For instance, a group of bones recovered from the Qafza shelter, close to Nazareth in Palestine, appear to be indistinguishable from the modern average. The dating indications available so far point to a period early in the Ice Age; furthermore, the remains were associated with an early Mousterian tool kit. This disconcerting find disposes forever of the convenient assumption that only "primitive Neanderthals" could be responsible for the relatively crude Mousterian flints. It suggests that the biological evolution of prehistoric peoples could have taken a leap forward without obvious signs of it appearing in their tastes for tools and weapons.

A number of other recent finds from Asia and Africa point strongly to the fact that fully modern characteristics were developing there at an unexpectedly early date, well before the beginning of the Ice Age. Later on, as the physique of the western Neanderthals stayed comparatively static, more progressive types of people appeared in the Near East and in central Europe. How exactly do we account for these facts?

It is intriguing that the eastern end of the Mediterranean was the setting not just for very early advances in physical anatomy but also for precocious developments in toolmaking as well. There was, of course, no "Eureka!" moment, when a Near Eastern hunter abruptly devised a range of bone and flint tools typical of the Upper Paleolithic period. Many of the basic techniques belonged firmly within the capabilities and customs of the Mousterian craftsmen. Tools continued to be made in several production stages, starting with the preparation of the raw nodule of flint, which was struck with a pebble to form a flat, roughly circular lump or core. By delicate knocks at the base of this tortoise-shaped mass, the prehistoric craftsman could obtain thin but remarkably rigid and durable flakes in a wide variety of shapes. A second method, especially suited to the production of long "penknife" blades, involved tapping the top edge of the core to detach a slender fragment from its side. In this way, the artisan could work his way around the core toward its center, obtaining many blades from a single lump of flint.

This was no hit-or-miss affair. The behavior of flint is predictable but requires expert handling if the friable mass is not to disintegrate into useless lumps or shatter into minute fragments. A blow angled too deeply or applied too brutally will result in a hopelessly large

Three contrasting skulls from the last Ice Age. At the left is a "classic" Neanderthal, La Ferrassie I. In the center is a skull from Qafza, Palestine, one of the earliest fully modern types known. At the right is a skull from the Upper Paleolithic site of Předmosti, Czechoslovakia, of entirely modern appearance but exceptionally rugged build.

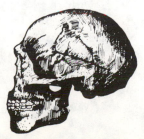

flake, quite impractical for the further effort of refining it into a tool. At too shallow an angle, the inexperienced flint knapper will shear off tiny and equally unworkable pieces. The Mousterians were certainly sufficiently skilled to produce these long blades, often in considerable numbers, but the consistent emphasis on their manufacture in many Upper Paleolithic tool kits was new. It meant a more efficient use of the raw material since many blades could be extracted from a single piece of flint. This in turn must have saved some of the time and energy necessary to find and gather suitable lumps from the flint beds along river valleys and streams.

From these common initial stages, the people of successive Paleolithic cultures fashioned a bewildering variety of finished tools in every conceivable style. Once again, this called for controlled and experienced hands. To create effective cutting or scraping edges, the craftsmen used "soft" implements of bone, wood, or horn to detach a row of minute, half-moon flakes along the sides or the end of the blade so that some tools appear to have been nibbled into shape— indeed, the term "nibbled retouch" is sometimes used to refer to it. Some of these flint blades were then fitted into wooden or bone handles, which have rarely survived.

None of these tricks of craftsmanship seem to have been the original invention of Upper Paleolithic communities. Yet the variety and ingenuity with which these skills were set to work contrast sharply with the conservatism of Mousterian traditions, in which tool types were repeated unimaginatively over thousands of years. In the earliest Upper Paleolithic (or "Aurignacian") cultures of France, for example, four completely new forms of flint implement are present, which probably reflect a wider range of practical skills in working wood and bone. Another innovation was the devising of multiple-purpose tools—blades with two worked edges instead of one—suitable for some combination of cutting, drilling, and scraping. In marked contrast to the Mousterian, every few thousand years of the Upper Paleolithic is marked by the introduction of newly invented forms and by some characteristic specialty of retouching that the experienced prehistorian can recognize at a glance.

Perhaps the most distinctive change of all in the equipment of the hunters was an entirely fresh range of carved objects in bone, antler, and ivory. Instead of the roughly worked splinters of bone that are occasionally encountered in Mousterian layers, the material was now carefully prepared and treated. The required lengths were obtained by two methods, either by splitting the bone lengthwise into small pieces with the help of a bone punch, or by carving twin parallel grooves with flint knives deeply enough so that long splinters could be prized out of position. These fragments were elaborately trimmed with flint tools and often were polished smooth against stone surfaces.

A range of Upper Paleolithic tools drawn from many different cultures and periods. Compare the diversity of shapes with the Mousterian tools shown on page 44.

Many of these technical developments occurred in different regions east and south of the Mediterranean long before they appeared elsewhere. Indeed, the trend toward blade tools began so early in areas like the Negev, the Nile Valley, and coastal Lebanon that the radiocarbon dating technique is useless (the method becomes increasingly unreliable after about 45,000 B.C.). From one or several such areas, the new tools spread progressively toward Europe. By 38,000 B.C. they had certainly reached Iran and the Balkans and probably central Europe as well.

The vital problem is to try and understand how and why these new fashions in toolmaking were adopted by one region after another. Some researchers, anxious to counter the crude idea of one vast migration from the eastern end of the Mediterranean, suggest that technical ideas could be transmitted from one community to the next, while the human populations actually "stayed put." In a similar vein, it is suggested that in certain regions, like central Europe, fully modern man could have evolved quite independently of physical developments elsewhere.

These both are distinct possibilities, but in fact, there *are* hints of major expansions of population at this period. One is that sites with the new Upper Paleolithic tools are much more numerous throughout Europe than those of their Mousterian predecessors. A more convincing point is that vast regions where the Mousterians never ventured were opened up and colonized for the first time. It is now known that blade tools appeared in northern Afghanistan well before 32,000 B.C., and from then on, the "frontiers" of northern and eastern Siberia were gradually pushed back. An important group of sites recently found on the banks of the Aldan river in eastern Siberia may date as far back as 33,000 B.C. The way was now open for the pioneering of all the great territory that lay from the fringes of the Siberian plateau right across to the Bering Straits and ultimately to the New World.

Piecing all this diverse evidence together into a recognizable picture is an exceedingly delicate task. Ever since Boule's day, it has been easy to imagine the spread of modern man and his industries in terms that are appropriate to nineteenth-century colonial empires, not to the huge time spans and sparse populations of prehistory. Fired by their newly found racial and technological superiority, "armies" of modern men can be envisaged setting off from Asia or the Near East, effortlessly vanquishing the Mousterians in their path.

The absurdity of this reconstruction is to some extent obvious. For one thing, we have recognized the possibility that fully modern man existed for many thousands of years in the Near East (and perhaps elsewhere), while still practicing the Mousterian styles of flint-working. So the great expansion of the new Upper Paleolithic tools from the Near East may have had little to do with human evolution in these regions or with the spread of a single race. Instead, it is more likely to have been a response to changing factors in the climate and environment. Just as the second phase of the Ice Age opened up great new tracts of "prairie" landscape for Mousterian settlement, so too the third cold climax appears to have extended favorable hunting lands right on into the farthest reaches of Siberia.

It is essential to bear in mind that these events took place over thousands of years, not decades or centuries. Would the people of the Upper Paleolithic communities actually be aware of their slow spread across these immense territories? Would they necessarily have wiped out the Mousterian peoples of Europe, rather than assimilating or coexisting with them over the generations?

There remains the obstinate question of the "classic" Neanderthals of the west, those abused and misunderstood figures who apparently never fitted into any pattern of physical evolution toward the form of modern man. What was the manner of their going? Was it with the tragic finality that has appealed to so many popular writers on the past (notably William Golding in his vivid novel *The Inheritors*)?

The "hard" evidence is so fragile and ambiguous that there can be no confident verdict on their fate. However, there *are* a number of European industries, like the Châtelperronian, that appear to be a strange mixture of Mousterian and more up-to-date techniques. One possibility is that these represent the handiwork of the last generations of the "classic" Neanderthals, who were influenced by the new artistic and technical skills of the incoming modern men.

There are now enough firmly dated sites to show that in parts of central Europe and southern France these peculiar hybrid styles persisted for an indefinite period of time (was it centuries or a few millennia?) alongside the earliest Upper Paleolithic traditions of the Aurignacian. For example, the Châtelperronians, who erected their mammoth bone huts at the Reindeer Cave at Arcy sometime around 31,000 B.C., were almost certainly contemporaries of the Aurignacians, whose arrival in the Vézère valley is timed at 32,000 or 31,000 B.C. It seems reasonable to suggest that the inhabitants of the Reindeer Cave acquired their taste for antler and bone points, decorative pendants, notched bones, and perhaps even domestic hygiene by some degree of peaceful contact with the new order. How long this lasted and whether it is right to imagine the Châtelperronians eventually succumbing to isolation or extermination remain, for the moment, unanswerable questions.

Earth Colors and Bone Music

The beginning of the Upper Paleolithic period still appears to represent a watershed in European prehistory. If the biological and social meaning of this watershed is obscure, the developments in material culture are nevertheless striking. They took the form of the technological changes incorporated in the Aurignacian tool kits already described. They included the first-known instances of the trading of such raw materials as flint and amber, which were sometimes passed across hundreds of kilometers from their place of origin (the full significance of this activity is discussed in Chapter 8). The most conspicuous of all the changes, however, was the appearance of many new forms of self-expression. Small sculptures and engravings of human and animal forms replace the few, faltering scratches of Mousterian times.

While certain of these new representational artworks have the rudimentary quality one might expect, others are highly accomplished. Among these is a handsome horse sculpture, one of a series of mammoth bone animals excavated from the Vogelherd cave near Württemberg, Germany, and dated to about 30,000 B.C. The horse, with its tense and arching neck, is heavily polished, as if from constant handling, and is in fact small enough to fit comfortably in the palm of the hand. The other animal statuettes from Vogelherd share the same expressive, elongated necklines and orderly arrangements of abstract signs, mainly dots and crosses, sometimes covering the entire body. In the case of one animal, probably a deer, the body was decorated with swift strokes after the head had, for some reason, been removed. These markings imply that the statuettes had some value beyond their undoubted artistic power.

In contrast to these and other remarkable sculptures from central Europe, the earliest works from the west consist mainly of unimpressive lines engraved on the surfaces of stone slabs, some of them accentuated with red or black coloring materials. It is clear that several of the blocks originally formed parts of larger compositions, presumably dislodged from the roofs and walls of the rock shelters by the action of weathering. A whole series of blocks came from Aurignacian deposits at the shelter of La Ferrassie, the site that also yielded the cup-marked Mousterian slab. The Aurignacian markings consist once again of cup-marks, together with oval outlines usually identified as vulvas. There also are painted and engraved animals, including a single obvious rhinoceros and several other less clearly defined creatures. On one comparable slab from the nearby shelter of the Abri Cellier, a vulva appears on the neck of an animal. These primitive designs appear to mark the birth of cave decorating practices, which were to continue for another 20,000 years in many varied forms. It is intriguing that, from the very start, animal representations seem to have been associated with abstract signs and symbols.

As well as decoration on a monumental scale, personal ornaments also seem to have been an innovation of the Upper Paleolithic. Necklaces made of pierced animal teeth, seashells, or attractive stones become common, while there is indirect evidence of body decoration. An excavator at Le Mas d'Azil in the Pyrenees, for example, uncovered a small, flat cake of red ocher, pitted with holes from which tiny amounts of the substance had been removed. Close by was a number of fine, sharp bone needles, and this led him to suggest that the Upper Paleolithic inhabitants of Le Mas d'Azil were engaged in tattooing.

As we have seen, less precise forms of body painting may have been common in Mousterian times. However, the finds of Upper Paleolithic ocher are distinguished from those of earlier epochs by the sheer volume of the coloring materials. It is not uncommon for the archaeologist, in the course of his painstaking, centimeter by centimeter, removal of a cave entrance deposit, to encounter an entire living floor vividly streaked and stained with red. At one Paleolithic site, the author assisted as a cache of black pigment was revealed and extracted. It was several centimeters thick and still greasy to the touch from a substance (animal fat?) that was originally blended into the powder as a binding agent. Whatever the exact purpose of these coloring materials, their appearance in such quantities does suggest a new level of artistic and perhaps ceremonial activity.

These actions may have been accompanied by the shrill pipings of the earliest European wind instruments to have survived from prehistoric times. Several varieties of "penny whistle" seem to have been current from the outset. The simplest type was made on a hollow reindeer toe bone, pierced with a single hole in the center, which was cross blown like a flute. Quite a number of small whistles have

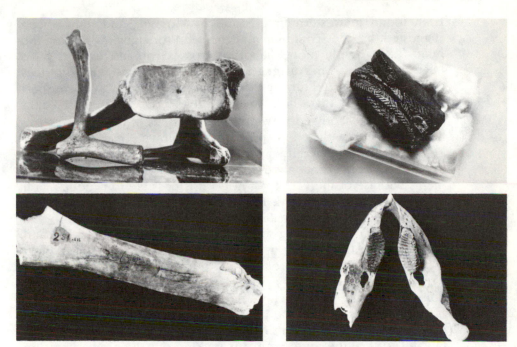

The original jam session. Assistants at the Ukrainian Museum of Archaeology, Kiev, are here seen in 1976 recreating the sounds of an Upper Paleolithic "orchestra" using painted mammoth bones found at the site of Mezin. The instruments consist of a mammoth shoulder blade as a drum *(top left)* with an antler hammer, a mammoth hip bone said to resonate with different tones like a xylophone, and *(right, top and bottom)* parts of a "castanet" and "rattler" made of mammoth jaws. The bones date from about 20,000 B.C.

been found in western France and in Moravia, and it has been suggested that they could have been useful for signaling during the hunt. Two slightly more elaborate cross-blown instruments have been recovered from Aurignacian deposits in the western Pyrenees and from Istállóskö, Hungary, both consisting of hollow bones perforated with two finger holes and a larger "blow hole": these could be held in one hand and played. Other versions of this basic type of flute, together with a few possible end-blown "pan pipes," are known from later Paleolithic periods. Unfortunately, it is impossible to reach any conclusions about the tonal ranges of these crude instruments because of the variety of notes it is possible to produce with different playing styles and, in some fragmentary cases, because of our ignorance of the original length of the tubes. Other bone objects may have emitted startling sounds, like the carefully engraved and ochered piece from a cave at Lalinde in the Dordogne. This object is identical to the "bull-roarers" that the Australian Aborigines whirled around on a string at different speeds to produce a booming noise of varying pitch. Finally, there would have been no shortage of percussive effects in the depths of the decorated caves, where every stalactite rings with its own bell-like note; perhaps it is not too fanciful to imagine the cave artists picking out their reverberating melodies, just as the tourist guides do today.

Exotic Tools and Invading Hordes

How can we reconstruct the lives of the people who were responsible for so many innovations at the start of the Upper Paleolithic period? Is it possible, for instance, to write a straightforward account of the events that followed over the succeeding millennia?

For many years, prehistorians have treated the evidence of flint tools like historical documents, "reading" into them the fluctuating fortunes of particular prehistoric hunting peoples. The appearance of a new tool kit was usually taken to herald the invasion of an alien race or group, while the possibilities of "home-grown" inventions or the influences of changes in climate and vegetation were neglected. The shortcomings of the simple "history book" approach have become increasingly obvious as precision in dating and in environmental methods has grown. Indeed, the intricate arguments surrounding the advent of modern man show how difficult it is to be sure of the identity of particular toolmaking groups. We can write several different versions of the demise of the Neanderthals in western Europe, depending on who we think was manufacturing the various tool industries involved in the drama.

This type of dilemma has confronted experts on the Upper Paleolithic period in no less bewildering a fashion, and sometimes their explanations for differences in tool kits have not lacked color or implausibility. A classic case concerns the industries that, in southern France, have been claimed to coexist with the Aurignacian culture throughout the first few thousand years of the Upper Paleolithic. The evidence of these distinct toolmaking trends was first presented by archaeologist Denis Peyrony in the 1930s, after many years of pioneering study of the cave deposits in the Eyzies region. Peyrony was convinced that the so-called Perigordian culture had flourished alongside the Aurignacian for centuries. He interpreted the two tool traditions as belonging to separate races that engaged in a full-scale war over possession of the rock shelters of the Dordogne. The relative frequency with which the two flint-working groups emerged in the archaeological records during the ensuing millennia was seen as a measure of the military successes of one side against the other, as first the Aurignacians and then the Perigordians scored their battle victories. A supporter of Peyrony's viewpoint, F. Lacorre, spun the fantasy further and equated the industries and peoples of Paleolithic Africa with those of southern France, and so provided an ultimate origin for the "negroid" Perigordian race in the valleys of the Indus and the Ganges.

Fifty years ago, when this chart first appeared, the past could be fitted into a simple and orderly sequence. Here the southwest French cultures are used as a framework for the entire world's prehistory. Each culture neatly follows the last, with no suggestion of an overlap in time. The starting point of the Upper Paleolithic is almost 20,000 years too late. For a simplified modern view of the French sequence alone, see the chart on page 90.

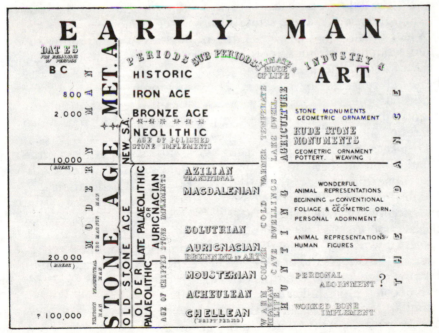

It is not surprising that, from the start, there were other prehistorians who were skeptical of such sweeping claims of racial and cultural identities. One of them was the Abbé Breuil, who in 1935 observed that:

> It is indeed most improbable that two peoples with completely different industries, the typical Aurignacian and the Perigordian, should have been able to live during the same era in two places as close as the Gorge d'Enfer and Laugerie-Haute.[4]

The two locations in question are a brisk ten-minute-walk apart along a modern road near Les Eyzies. Could an intertribal war have raged for thousands of years between neighboring sites?

Today, more than forty years since the theory of contemporary "Aurignacians" and "Perigordians" was born, the debate among French prehistorians is still far from resolved. Many experts still cling to the idea of distinct ethnic groups "rubbing shoulders" for thousands of years in the river valleys of the southwest. If there really were such distinct peoples during Aurignacian times, then the Perigordians were certainly few and far between. It is only much later, around 26,000 B.C., that the Aurignacian sites diminish in number and the Perigordian traditions take over most of the celebrated French cave sites. This change was presumably not the result of some flint-and-bone Waterloo gained at the expense of the Aurignacians. Fictional explanations of this kind must be weighed against more prosaic possibilities: perhaps there was simply a change in the tool-making preferences of the same groups of craftsmen.

Whatever the right answer is, the difficulties in interpreting detailed sequences of stone tools in particular regions are clear. Do these represent the work of one group of people or several? Are there differences to be observed because sites were occupied for varied purposes, perhaps at separate seasons of the year? It is only recently that these sorts of questions were even considered, much less tested against the fragmentary evidence available.

To take another example, recent work shows that at least three variations on the Perigordian style flourished side by side in southwest France during the period around 25,000 B.C. A generation ago, a prehistorian would have happily ascribed these varieties to the influence of separate tribes. Today, at least one eminent French scholar supports the idea that different activities or preferences may be seen at work within a single Perigordian people. This new recognition of the complexity of the evidence explains why, as the weight of knowledge grows, local histories of Upper Paleolithic peoples have become more difficult to write, not less.

Similar intriguing developments also have begun to emerge from Ice Age research in the Soviet Union. For many decades, it has been recognized that the hunters of southwestern Russia, who raised their homes of hide and mammoth bones, turned out their flint tools in styles broadly similar to the Perigordian (also known as "Gravettian") traditions farther to the west. For several thousand years—longer in some places than others—there seems to have existed a basic cultural unity between the peoples who occupied the enormous area all the way from the Don Basin in the extreme east right across to the shores of the Atlantic. One of the habits that seems to link these diverse populations was that of making little bone or ivory human sculptures, known rather misleadingly as "Venus" figurines. These include some of the most celebrated female images known in prehistoric art, such as the statuettes from Willendorf and Lespugue, although whether or not they all express a single meaning or theme is a difficult problem (discussed in Chapter 11). At any rate, these very widely scattered peoples appear to have had certain common attitudes or tastes that found expression in their flint and bonework. At the same time, the new Soviet research shows a hitherto unexpected level of complexity present within the Russian material.

One of the most remarkable recent finds took place in the suburbs of Vladimir near Moscow, at a site known as Sungir. It consisted of a large Upper Paleolithic settlement and a cemetery with a number of well-preserved burials, dating to about 23,000 B.C. The most spectacular grave, revealed in 1969, contained the bodies of two boys, one aged seven to nine, the other twelve to thirteen, laid head-to-head in the trench. This burial was removed from the ground in one great block weighing several tons and was taken to a Moscow laboratory, where it was painstakingly examined. The skeletons of the boys were covered with thousands of pierced mammoth ivory beads, which were once sewn onto their clothing as ornaments. The boys wore headdresses decorated with these beads, together with the canine teeth of polar foxes. There were embellishments in the form of bracelets, rings, chest plaques, and other objects, all of bone, while the boys were equipped with a formidable armory of sixteen spears, darts, and daggers. Several of these spears were heavy objects, magnificently worked from mammoth ivory to a length of over two meters. It is not known how the Paleolithic craftsmen managed to straighten the natural curvature of mammoth tusks to produce these remarkable weapons, which are unique in the prehistoric record.

The Sungir burials were the most striking finds to be associated with a type of flint-working entirely different in character to the Perigordian or Gravettian style. The discoveries of the Sungir-type tools now extend from the area of the Black Sea right up to the Arctic Ocean, and in some regions, these sites undoubtedly represent the earliest known appearances of the Upper Paleolithic. However, in southwest Russia, the Sungir industries can be seen to overlap with

The first of the Sungir burials as it was unearthed in 1964, preserved below the permafrost in a grave stained bright red with ocher. The burial was of an elderly man clad in lavishly beaded fur clothing dating to at least 23,000 B.C.

the Gravettian culture both in time and in space. Once again, we face the classic questions: do these tools reflect rival bands of "Gravettians" and "Sungirians" competing for what must have been the richest hunting territory in western Russia? Or are events of a more complicated and less dramatic nature involved? In this case, the character of the Sungir-type tools is so distinctive that a separate ethnic group may well be the answer.

Despite the great advances that have been made in the practical methods of dating Ice Age sites and reconstructing their environments, an understanding of what these contrasting patterns of tools and bones actually represent is still in its infancy. The fault lies partly in the immense gaps that remain in the detailed record of particular regions and periods. Nor is the progress of this basic research in central and eastern Europe any less important than the study of the better known sequences of cultures in southern France, which have benefited from more than a century of scholarly and public attention. It will be recalled that many important developments during the Ice Age seem to have happened in central Europe long before they occurred in southern France. The hunters of Hungary and Moravia certainly adopted blade tools several thousand years before they appeared in France. This area also could have played a part in the physical evolution of modern man and in the spread of certain cultural fashions like the Venus figurines. In this light, the people who occupied the famous caves in the Dordogne and in northern Spain actually seem like "provincials," in a far western "backwater," remote from influential changes that took place in central Europe. Indeed, toward the end of the Ice Age, the people of this "backwater" seem to have gone their own way in the development of working flint and bone. At about 21,000 B.C., the Perigordian traditions disappear in the west, and their place is taken by two brilliant local

View of the most spectacular discovery at Sungir in 1969, the burial of two boys laid head-to-head with a rich array of beaded clothing, ornaments and weapons.

developments, first the Solutrean (about 20,000–16,000 B.C.) and then the Magdalenian (16,000–9,000 B.C.). Both these cultures were concentrated in southern France and in parts of Spain, while the inhabitants of the rest of Europe carried on their ancient customs of craftsmanship, largely unaffected by events in the west.

It is with the Magdalenian, in particular, that the great master-pieces of cave painting and engraving are to be associated. The question of why these unsurpassed artistic achievements occurred mainly in southern France and northern Spain is related intimately to the history and origins of the Magdalenian people. The products of their artists are set apart, both in style and quality, from anything comparable in the rest of Europe at the end of the Ice Age. Caves *were* decorated in many regions outside the Magdalenian influence, for example, in the canyons of Languedoc and along the rocky coast-lines of Sicily, but the vision of the Mediterranean artists was differ-ent. They followed archaic conventions of style that resulted in rather stiff and formal animal designs, quite unlike the vigorous, spontaneous images of the Magdalenian painters of Lascaux and Altamira.

What encouraged this separate path of development in the west —the cultural "hothouse" of France and Spain—in the first place? Wild and exotic theories have been advanced to account for the origin of the Solutrean culture, largely because the remains of this culture are associated with a remarkable type of flint weapon.

These leaf-shaped spearheads, sometimes as thin and as brittle as glass, have been admired for their singular and unequaled craftsman-ship. The striking standard of finish was achieved by the expert control of a technique new to the flint workers of the west. It in-volved pressing down hard on a broad blade with the help of a bone or antler point, and so prizing shallow, horizontal flakes off its sur-face. This treatment was at first confined to only one side of the spearhead, but in the later stages of the Solutrean, the surface of the flint was entirely covered with horizontal ripples, the delicate fluted impressions left by a mosaic of detached flakes, combining together in a harmonious and symmetrical outline. Besides these celebrated "laurel leaves," a variety of other projectile types was produced by similar methods in the course of the Solutrean's brief appearance. Some of the spearheads are so wafer-thin that they would undoubt-edly snap if any pressure were applied to them, and these must surely represent items of prestige or exchange rather than practical hunting weapons.

The comparatively sudden emergence and disappearance of the unique spearheads inevitably encouraged speculation. Their limited distribution in central and southern France and a few fringe areas in Spain suggested nothing less than invasion. The Solutrean tribes had either migrated from northern Africa through Spain or had developed from the earlier cultures of central Europe and had been driven toward the west by a deteriorating climate. For some years, discoveries of carved or painted objects in Solutrean layers were rare, and this seemed to emphasize the exotic character of the Solutrean people. Indeed, in some textbooks of the 1920s and 1930s, the Solutreans are represented as a regressive, dull, and taciturn people, uninterested in the fine arts, but instead concerned only with the purposeless perfection of their flint-work. In fact, subsequent discoveries of superb sculptured reliefs and cave engravings of the Solutrean period suggest that there was probably no decline in artistic talent or break in the continuity of cave art.

Moreover, renewed studies of the Solutrean tools showed how traditional their crafts were, with the solitary exception of the spearheads. Many of the commoner types of flint tool closely follow earlier patterns, and only one bone-working innovation can be credited to them, the eyed sewing needle, so that we can now imagine Paleolithic people in skin garments more neatly fitted or embroidered than they could ever have been before.

In any case, the proof that the Solutrean spearhead was an original development somewhere within southern France is clear from new excavations in the rock shelters of the Dordogne. In the layers of these classic sites, every stage in their evolution can be traced, from the comparatively crude beginnings of the flat, single-faced type to the final, superbly refined, narrow points. It is obvious that the outlying pockets of this culture, notably along the margins of north-

Excavations in progress in 1976 in the Quercy region of southwest France at the Upper Paleolithic site of Les Fieux which lies at the entrance to a decorated cave.

ern, western, and eastern Spain, represent its last stages and not the trail of some imaginary African invasion. There is good evidence to support the idea of a steady rise in population throughout the Solutrean period, which resulted in the eventual spilling over into the coastal fringes of Spain.

The reinterpretation of the Solutrean culture was one of the most decisive proofs of the inventiveness and individuality of the late Paleolithic people of the west. The origin of the Magdalenian style of craftsmanship is not so well documented, but there is no obvious reason to invoke the arrival of aliens to account for its distinctiveness. Again, many features of its flint- and bone-working are recognizable as continuations of much older Ice Age traditions.

By now, the reader, weary of the constant succession of Paleolithic cultures, may be wondering how these textbook classifications can provide a real insight into the way of life led by the prehistoric hunters. Even if it were possible to write a confident "history book" account of the cultures, would this convey a vivid impression of what it was like to be alive during the Upper Paleolithic?

The People of the Horse

There are at least two ways of filling in the picture further. One is to present the economic, social, and environmental evidence gathered by the archaeologists and then to compare and contrast it with the records of recent hunter-gatherers like the Bushmen or the Australian Aborigines. Do the observations of nineteenth- and twentieth-century anthropologists help or hinder our attempts to reconstruct the remote past? This is one of the key problems considered in the following chapters. Another approach is simply to describe a single Magdalenian site in detail—not necessarily as typical of that culture or era—but in order to engage the imagination with the particular rather than with the general.

A remarkable campaign of research has revealed a great deal about one Magdalenian settlement called Duruthy, located in a commanding position on the fringes of the French Basque country. The river systems of the extreme southwest here converge on the great estuary of the Adour, which in its lower reaches formed a broad, boggy, and impassable barrier to traffic until recent times. The first convenient crossing point of the river system in prehistory must have been in the vicinity of Duruthy, and it is not surprising that this area of the Oloron valley shows signs of intensive activity at almost every period.

From the top of the Pastou cliff, high above Duruthy, one looks out to the south over the curve of the river, the tumbling green pasture of the Basque hills, and beyond to the high crests of the

Pyrenees, often lost in cloud or haze. At this excellent lookout point, the captain of the local garrison at Dax discovered traces of Iron Age fortifications in 1872, including a gold bracelet inside one of the camps. This encouraged him to search the area thoroughly, and he came across the rubbish deposits of four prehistoric settlements, strung out in a line at the foot of the Pastou cliff, including the site of Duruthy. Two years later, the pioneer prehistorians Louis Lartet and H. Chaplain-Duparc dug into part of the heap of debris at Duruthy and found a late Paleolithic burial accompanied by a necklace of fifty pierced lion and bear teeth, most of them engraved with curious abstract signs and the occasional animal motif. Duruthy was subsequently neglected until 1957, when archaeologist Robert Arambourou visited the site and, among other stray finds, picked up fragments of a human jaw lying on the surface. He was so impressed with the debris that he acquired permission to dig the site, and every year up to the present, he has devoted intense, and at times solitary, efforts to his investigation of the site. In recent years, the contributions of experts in soils and animal bones have greatly added to the interest of Arambourou's painstaking work.

There may be as much as 30,000 years of Ice Age occupation represented by the layers at Duruthy, but so far only the settlements of the Magdalenians have been revealed, covering the last 4,000 or 5,000 years of this period. The record begins in about 13,000 B.C., when the visits of the Magdalenians to the site were apparently of brief duration, perhaps no more than temporary halts in the course of hunting expeditions. This was during a period of intensely dry and cold conditions, when only a few pines grew in the valley. Herds of wild horses must have grazed in this open landscape, for they were the chief quarry of the Duruthy hunters. They may have been discouraged from sheltering too regularly under the shallow overhang because of the frequent avalanches of boulders, fractured from the cliff face by frost and ice.

From about 12,000 B.C., a warmer and more humid climate prevailed; the pine trees multiplied, and they were joined by other such species as alder, hazel, birch, and willow. For a period of 500 years or so, settlement continued at Duruthy, but it was punctuated by a dozen long intervals when the hunters were apparently installed elsewhere. During their stays, the people of the Magdalenian IV culture now mainly hunted bison, and later on horse and reindeer became important. Discarded seashells and a few teeth of dolphin were presumably "souvenirs" collected at the nearby coast. The people of Duruthy were now also making or acquiring exquisitely decorated art objects of bone and stone.

In 1961, Arambourou made an extraordinary discovery in the levels of this middle Magdalenian phase. Close to the cliff, he uncovered, all within the space of a single meter, a rich collection of art objects. There was a fine horse head carved from white limestone and pierced as if it were intended to be worn as a pendant. A second, less expressive horse head of ivory was found close by, and this appeared to have been treated in a more abstract style or else left unfinished by its creator. A few centimeters from the first head was a fragile little round bead of jet, which disintegrated as attempts were made to remove it.

But the third sculpture recovered at Duruthy exceeds all these finds in its quality and interest. It is the largest portable object of Paleolithic art ever found and one of the most striking. Fashioned from a sandstone block some twenty-seven centimeters long, the bold head and shoulders of a horse are thrust forward, while the details of the rest of its body are not clear. The block is too short to have accommodated the horse's legs, and it may be that they were intended to be shown as if folded under the body in an attitude of repose. Whatever the case, the disturbing power of this work is not easily forgotten by those who see it today. M. Arambourou expresses it perfectly:

> One cannot fail to be struck by the entirely hieratical character of this horse, which owes as much to the posture given to the animal as to the manner in which it is executed. With a realism which nevertheless suggests the particular rather than the gen-

The horse pendant of white limestone discovered at Duruthy, western Pyrenees, in 1961.

The magnificent horse sculpture found at Duruthy, the largest portable Paleolithic art object known.

eral, the sculptor has, it would seem, figured not a horse, but *the* horse. . . .[5]

Arambourou's speculation is supported by the fact that the horse was found lying on top of the jaw bones of two horses and up against two horse skulls. The association of all the objects found at this spot, with no trace of any human burial to accompany them, surely implies the existence of some kind of horse ritual or shrine of the middle Magdalenians.

The interest of Duruthy does not stop with this period of occupation by gifted hunter-artists. A cold climate once again returned to the southwest, and the settlements of the Magdalenian V people appear to have been meager and brief. When more favorable conditions returned, by about 9,500 B.C., Duruthy was occupied by the last of the Magdalenians on a scale larger than before. Over three times the area of the slope in front of the cliff was occupied than had been settled in the days of the middle Magdalenians. This area was consolidated by great artificial pavements laid down with river cobbles in a series of terraces, perhaps to even out the slope or to make it less slippery. The pavements were rebuilt eight times by the late Magdalenians; the earliest covered the burial of a woman in her fifties, while the latest preserved the traces of post-holes on its top surface, presumably to support tentlike structures.

Together with the other communities installed at the four neighboring sites now known from the foot of the Pastou cliff, it seems that the population of the Oloron valley must have lived, as Arambourou puts it, in "a sort of large village." However, the occupation of Duruthy was always seasonal, during the months of September to February only, as studies of reindeer teeth have revealed. In autumn, the hunters caught salmon with bone harpoons (the backbones of the fish are constantly found with these weapons), and during the winter, reindeer was the staple. In the summertime, the people of the Oloron "village" may well have dispersed in order to follow the reindeer herds as they migrated up to the cool mountain pastures.

The last Magdalenians seem to have been uninterested in the arts (although ocher crayons and pierced teeth have been found), but the quality and inventiveness of their flint-work was unsurpassed. It is difficult to convey any impression of the prolific output of the Duruthy toolmakers or of the skill and fineness applied to the basic forms; 10,000 worked tools have been studied from the Magdalenian VI layers, a large proportion of which are tiny backed blades worked on an almost invisible scale. When these bladelets were examined under a microscope, it was found that they bore traces of glue on the blunted backs, and a peculiar sheen on the cutting edge that can only have come from the cutting of grass stems. It seems likely, then, that the bladelets were cemented into a wooden or bone handle to form sickles that were used to harvest the wild grasses in the Oloron valley. Arambourou also has identified the remains of what may be stone mortars and grinders, which would have been essential for turning the grasses into palatable flour. This remarkable evidence suggests that some kind of bread, porridge, or cake, perhaps cooked on hot stones near the fire, accompanied the reindeer feasts held in the shelter of the Pastou cliff. It seems as if the brilliance of the Magdalenian IV artists was matched by the technical ingenuity of their Magdalenian VI successors.

The Duruthy research shows how much can be revealed by the patient and systematic investigation of a single site. Can this type of information, drawn from many sources, be used to build an image of the hunting and gathering life, both in the remote past and in the present?

Engraved bear tooth from the burial discovered at Duruthy in 1874.

5: *The Question of Survival*

A Lost World

How readily the imagination creates an inhospitable world for pre-historic people. The term "Ice Age" itself summons up empty land-scapes, bleak deserts of moss and stunted shrubs—the Siberia de-scribed by travelers a century ago:

> The peculiar bluish hue of the tundra, and the vast expanse of its flat surface, present to the traveller a curious illusion of having before him a great waste of waters rather than a plain. This resemblance to the sea is heightened when moonlight floods the tundra, or when the wind has heaped up a light snowfall into dunes and undulating furrows.[1]

For the lonely human figures lost in this prehistoric wilderness, there were surely few comforts in the freezing winds, the clinging damp of tent and cave walls, and, above all, the compulsion to wander ceaselessly in the tracks of the restless reindeer. How can we recon-cile this insecure and dispiriting world, the Ice Age of so much popular writing, with the great size of a settlement like Duruthy or the splendid vision of paintings like those at Altamira?

The answer lies partly in defects in the popular picture, especially in the assumption that prehistoric hunters lived in environments as hostile as the Arctic wastes where scattered remains of the traditional Eskimo way of life persist today. The one clear fact to emerge from the last two decades of scientific study is that no modern landscape or climate reproduces the circumstances in which Ice Age communi-ties found themselves. It was a unique environment, which can be matched in parts but never in its entirety with any modern habitat. For this reason, reconstructions of the prehistoric landscape have the fascination of a "lost world," while they also raise difficulties in understanding exactly how animals and humans responded to and influenced their surroundings.

An attempt to summarize a "typical" Ice Age environment would be as absurd as trying to describe a "typical" countryside or climate of modern Europe. As the work of experts in the study of pollens and sediments has increasingly shown, there were not only variations between and within regions, just like today, but the character of these contrasts was sometimes rather different from that of the pre-sent.

The bleak Ice Age landscape of popular imagination: wet high tundra photographed in northern Finland.

Consider, for example, the evidence from two regions where a few sites have been studied with sufficient care to provoke interesting comparisons: the Périgord region of southwest France near the Atlantic, and the area of the Midi in the southeast, stretching down to the shores of the Mediterranean. During one phase of the Ice Age, from about 55,000 to 35,000 years ago, the evidence suggests that most of southern France was subjected to increasingly dry and cold conditions. Animals that must have flourished in open grasslands, such as horses and large bovids, become increasingly prominent in the lists of bone remains from key sites. Yet this general climatic change was not a smooth process, for it was accompanied by five especially sharp peaks of cold, revealed in archaeological layers mainly by the presence of large blocks, split and shattered by severe frosts from cave walls and ceilings. The effects of the climatic change also were different in the two major regions of the southwest and southeast. One piece of evidence for this is that reindeer dominate the lists of species for the southwest, while in the southeast, their bones generally appear only west of the Rhône valley. Just like today, there was a contrast between the more rigorous, continental climate of the southwest and the warmer, milder conditions of the southeast. Quite unlike the present day, however, the evidence also suggests that there was much more humidity in the Midi region. Here, preserved in the cave of L'Hortus, the remnants of pollen, large mammals, rodents, grasshoppers, and birds have all been studied and combine to present a picture of a climate drier and colder

In reality, during mild intervals of climate, much of northern Europe must have looked similar to this scene, lightly wooded with birch and spruce, in central Sweden.

than now. Despite this, the consistent appearance of trees and shrubs of a typically Mediterranean form and favored only rarely elsewhere suggests that, 50,000 years ago, as now, the south of France was one of the most agreeable spots in Europe.

This glance at just one phase of the Ice Age in only two regions demonstrates something of the complexity that faces the experts who try to reconstruct the landscapes and climates of the period. As we follow their studies through time, the complications increase rather than diminish. As more and more sites appear toward the end of the Ice Age, it becomes possible to measure changes in climate and vegetation in sharper detail. For example, a sample of soil extracted from a layer on which the early hunters of the Aurignacian period once camped can be taken to a laboratory and usually identified to within 3,000 or 4,000 years of a particular climatic phase. At the other end of the scale in Magdalenian times, a similar soil sample can be placed to within 1,000 years of the right date, while pollen scientists are currently debating the nature and duration of climatic phases as little as two or three centuries long. Each of these fluctuations had its own particular regional impact on vegetation and on animal and human populations; yet simply tracing the sequences and changes without trying to understand them is difficult enough. Just as it is hard to imagine the passing of thousands upon thousands of years in the archaeological layers, so, too, is it difficult to visualize the innumerable swings between open and forested landscapes, between arid and humid climates.

Two contrasting environments from the Mousterian period (during Würm
II) reconstructed from the study of the Hortus Cave in southeast France
(situated at the foot of the cliff in the background). At the stage seen above,
pine trees were replacing oak as more arid conditions set in. At a later phase
(below), there was an open prairie landscape, while the less hardy vegetation
retreated to the sheltered slopes below the cave.

One fact stands out in the midst of all the complexity: the true high tundra so often associated with prehistoric hunters was rarely present in western Europe even in the most bitter climatic periods. Despite the undoubted sharpness and length of winters during cold phases, there are many reasons for supposing that Paleolithic environments were always richer than those in which modern Eskimos live. In many regions of central Europe, the ancient hunters were camping not on the waterlogged, acidic wastes of the Arctic, but on fertile loess soils windblown from the great glacial deposits. The soils undoubtedly favored the spread of vast expanses of grasslands. Furthermore, these open plains were not obscured by the endless nights of the Arctic regions, but because of their more southerly location, they enjoyed more direct solar radiation. This in turn favored a diversity of plant life and grazing possibilities unknown in the extreme northern latitudes, but comparable, perhaps, to the steppes of European Russia, which, as late as the 1850s, supported herds of wild horses, many of them several hundred strong.

However barren these landscapes were, the evidence from a great many living sites in western Europe typically shows some traces of tree pollen even during the most severe phases. This suggests that the grassy steppes were often interrupted by wooded areas, no doubt clustered around river valleys and sheltered hillsides. In fact, one of the more interesting points to emerge from recent studies is that prehistoric hunters in Europe often chose sheltered spots to settle where especially favorable local conditions of climate existed. Analyses of these sites show a much greater variety of plant and animal life than existed on the open plateaus nearby.

Two places in western Europe illustrate this particularly well, one a small, sheltered valley at La Marche in southwest France, the other at Gönnersdorf, overlooking the banks of the Rhine in Germany. Magdalenian hunters selected both of these locations for settlements or sanctuaries (described in detail in Chapter 12). In the valley of La Marche, the abundance of pollen from lime trees and hazel, which require temperate conditions, is remarkable, considering that the settlement itself contained many discarded reindeer bones. This indicates that the surrounding upland country was probably open grassland. Meanwhile, the hunters at Gönnersdorf, protected from bitter north winds by the hillside on which they had settled, looked out over a Rhine valley thick with pine and other species, despite the fact that pollen samples from the neighboring plateau show no traces of trees, except for a few birches. It is obvious that prehistoric hunters would have chosen favorable spots for their living bases, but the existence of such extreme contrasts in the local vegetation is surprising.

More often than not, then, the Paleolithic hunters seem to have occupied a patchwork of grassy plains and wooded valleys, a mosaic

A chart summarizing crudely what is known of the sequence of climates and cultures in southwest France.

CLIMATIC PHASES		RADIO-CARBON DATES OF PHASES	RECONSTRUCTED CLIMATE			SEQUENCE OF ARCHAEOLOGICAL CULTURES	
			HOT	COLD	DRY HUMID		
POST-GLACIAL	Subatlantic	2700 ±50				METALS	
	Subboreal	4500 ±300				NEOLITHIC	
	Atlantic						
	Boreal	7500 ± 100					Post-Glacial Cultures
	Preboreal	9000 ± 200				MESOLITHIC	
WÜRM IV	Dryas III	10200 ± 200					Azilian
	Alleröd	10800 ± 200					VI
	Dryas II	11800 ± 200					V
	Bölling	12300 ± 200					IV
	Dryas Ic	13300 ± 200					Magdalenian III
	Prebölling						II
	Dryas Ib						
	Lascaux						
	Dryas Ia	17000					I
	Laugerie						
WÜRM III	Würm IIIc	19000 20000				UPPER PALEOLITHIC	Solutrean
	Tursac	24000					
	Würm IIIb	28000					Perigordian (Gravettian)
	Salpetrière						
	Würm IIIa						
	Arcy	32000					Aurignacian
	Würm IIIa						Châtelperronian
	Hengelo-Quinson	36000				M. PAL.	Final Mousterian

landscape, that encouraged a remarkable diversity of animal species. Indeed, side by side in the archaeological layers we find the bones of creatures that never occupy the same habitats in the modern world, for example, the reindeer of the Arctic barrens and the saïga antelope of the rocky Alpine foothills. Unexpected associations like this are found right across Europe, encompassing a vast range of animal life. In the dense occupation sites of the Russian Ukraine, the presence of snow lemmings indicates an environment colder than today; yet other rodents, such as the great jerboa, are typical of temperate grasslands, and the modern distribution of the two animals never overlaps. This emphasizes, again, the unique character of the Ice Age world. The survival strategies of beleaguered Arctic Eskimos give us no direct ideas of how prehistoric hunters lived, faced by grasslands apparently teeming with reindeer, horses, and mammoth; we cannot translate the entire pattern of their experience and skills back into the vanished Ice Age.

The Ice Age Diet

However, lessons of a more limited kind still await the investigator prepared to compare ancient and modern hunters. Too often anthropologists, intent on observing patterns of social behavior, have simply not troubled to record the kind of detailed information that would be useful to an archaeologist. Very few investigators have compiled detailed plans of campsites or have attempted to see whether or not certain activities leave behind distinctive types of refuse in the ground. If we find a scatter of bones at a Paleolithic site that can be compared to that of a Bushman camp, it does not mean that we should imagine the environment of the Kalahari desert in Ice Age Europe, but merely that the two groups are likely to have pursued certain activities in common.

A brilliant example of this neglected type of comparison was published by two French archaeologists in 1974. One of them, Jean-Philippe Rigaud, spent several months among the Nuniamut Eskimos of northern Alaska and excavated their abandoned living sites. One of his most interesting observations is that a notable contribution to the Nuniamut diet is made by the highly nutritious grease extracted from the middle of caribou bones. This involves three distinct methods of preparation: the marrow is extracted by smashing the long bones and is eaten cooked; soup is then made by boiling up small fragments of the same bones; finally, grease is extracted by breaking the ends of the bones into even smaller pieces and boiling them up for a long period. Each of these activities involves bone fragments of a distinctive size. Assuming that we could find an exactly similar pattern of breakages at a Paleolithic site, it would be reasonable to conclude that prehistoric hunters also practiced these same techniques.

Returning to his excavation at Le Flageolet in the Dordogne, Rigaud and his colleague, Delpech, were indeed able to recognize distinct groups of bone fragments identical to those produced by the Eskimos some 25,000 years later. These scatters were grouped around hearths containing burned stones, which were almost certainly used to boil water in containers (made of skins?) or to roast meat. It was even possible to demonstrate that the Perigordian hunters of Flageolet preferred the same parts of the skeleton for the extraction of marrow as the Nuniamut. Before Rigaud and Delpech's work, it was well known that large bones were split for their marrow (even the bones at the 2-million-year-old campsites of Olduvai Gorge show the marks of this). But Mousterian hunters as late as 40,000 years ago were apparently content just to smash the long bones without boiling up smaller fragments. The preparation of soup and grease seems to belong entirely to the Upper Paleolithic period, marking a step forward in the efficient use of animal resources. While this may not represent a major revelation of the prehistoric diet, nevertheless the clear and constructive use of a modern hunting group's activities in understanding those of the past is remarkable.

Normally, the example of recent hunter-gatherers is taken to show, in a far less positive spirit, the unreliability of our ideas of ancient diets, when these are based solely on scatters of animal bones. One case is the Hadza of Tanzania, who, until forced to adopt agriculture during the 1960s, pursued a traditional hunting-and-gathering economy near Lake Eyasi in the African rift valley. At this stage, they were studied by British anthropologist James Woodburn, who compiled careful accounts of their subsistence activities. Among the Hadza, the proportion of vegetables in their diet was high, as much as 80 percent. Much of this was gathered in daily excursions by the women and children, who satisfied their hunger on the spot and brought a quite small proportion back to camp. Despite the importance of vegetables, Hadza men thought of and described themselves as hunters, attaching both gastronomic and ritual significance to eating meat. Their camps were sometimes situated on high points commanding a good view of prospective game, but nothing like a constant surveillance was maintained. Instead, the men sat around gambling for much of the day until an individual sighted and pursued an animal. If he was successful in bringing down a small animal, he might eat it immediately or, if too large, invite his fellows to consume part and drag the remains back to camp. For this reason, limbs as well as complete carcasses were often carried home. Once the meat was cooked, it was divided according to age and sex, and in their large summer camps of the dry season, the men always ate separately from women and children.

What evidence would one of these large Hadza campsites leave behind for archaeologists? The animal bones would represent mainly the large species, so there would be no evidence of the small game consumed in the field. Such an unrepresentative selection of bones could prove even more biased if scavengers came and removed bones from the site just after the Hadza group had moved to another location. Furthermore, small-sized bones are much more vulnerable to processes of weathering and decomposition, so again the archaeologist might receive a false impression of the importance of big game animals. Finally, the sexually segregated eating habits of the Hadza might mislead the archaeologist still further, particularly if only a part of the campsite were uncovered.

A more crucial problem is the difficulty of estimating the importance of vegetables in the diet. Unless there were quite exceptional conditions of preservation on the site, no traces of these would be recovered by the ordinary techniques of excavation. The problem is basic to our understanding of the stresses and insecurities of the hunter-gatherer life because the demands of largely meat-free economies are very different from the meat-dependent. For hunting communities living in tropical latitudes and surrounded by luxuriant vegetable growth, protein requirements are often satisfied by as little as one hundred grams or less of meat a day. But if vegetables are scarce, as in the Arctic, and meat is required to provide calories as well as protein, then no less than twenty times this minimum amount of meat must be consumed in a day. In a cold climate, such a dependence makes good sense because of the body heat generated by the greater digestive effort involved in meat consumption. Yet the two extremes of diet demand very different ways of life, and it is natural to assume that a people dependent on the daily luck and skill of the hunter is more precariously poised than a group that can always fall back on a menu of roots, nuts, leaves, and berries if all else fails.

All these generalizations arise from the modern world of recent hunter-gatherers, where the importance of vegetables in the diet does indeed seem to vary with latitude, starting from about 80 percent or more at the tropics and falling to 5 percent or less in the Arctic (among the Nuniamut, who now eke out their existence with tinned foods as well as hunting, consumption of vegetables is said to have fallen to 1 percent). While we have no proof that this same pattern can be projected into prehistory, it seems reasonable to argue that the unique environment of Ice Age Europe was unlike either these tropical or Arctic extremes. From study of the pollen at sites with particularly favorable local climates, such as La Marche or Gönnersdorf, it is certain that edible roots and fruit resources were available, and it is unlikely that they were neglected by Paleolithic people. One prehistorian, discussing pollen traces preserved in a Pyrenean cave, commented:

We will probably never know if the inhabitants of the cave of Isturitz had a taste for willow buds or dandelion leaves, but we *do* know that these could be gathered in the vicinity of their living site.[2]

Direct proof of such activities would take the form of discarded fruit stones or nutshells dug up in the archaeological layers. Unfortunately, such remains are almost unknown for the Ice Age period, undoubtedly because of the slim chances of their preservation in extremes of coldness and aridity.

A more promising line of inquiry involves the study of tiny scratches left on the surfaces of human teeth by biting and chewing. This novel technique, still in its experimental stages, consists of the microscopic examination of the distinct patterns left by different diets. A largely vegetarian diet requires more mastication and leaves a different pattern of wear than one that is rich in meat. Only two examples of Paleolithic tooth wear have so far been examined in this way, but interestingly, their patterns of scratches seem to correspond to neither a wholly vegetarian nor meat-dependent regime, but to something in between.

So what little evidence there is indicates that many Ice Age communities in western Europe lived not in bleak and marginal surroundings, but that a wide range of animal and vegetable foods were there to be exploited throughout the year. Assuming that these guesses are correct, a vital question poses itself, one that affects our entire outlook on the stability and security of Paleolithic life: why should the hunters have chased one species of animal in preference to another? Why, in particular, did some hunting groups become so dependent on reindeer that the glacial period is often popularly known as the "Reindeer Age"?

Crisis and Security

Clear choices were already made by hunters living in the very earliest centuries of the Ice Age. As mild climates imperceptibly gave way to long, sharp winters and arid summers, people in central Europe 100,000 years ago deliberately selected the animal species that they pursued. Imagine a hunting station at that remote period, situated high over the Danube Valley at a spot now overlooking the suburbs of Budapest. Here, on the edge of the plateau of Érd, halfway between the mountains and the river, a group of hunters camped in two basins in the limestone, two little cavities side by side that both gradually filled up with soil, charcoal, animal bones, and Mousterian flint tools. The rock walls of the two cavities protected these occupation remains from being washed away and helped to preserve them to the present day.

The earliest human groups to settle at Érd, no doubt attracted by the shelter of the limestone walls, lived at the edge of a hill slope dominated by pine trees that, as the centuries wore on and the climate became more rigorous, gave way increasingly to larch. At the peak of one cold phase, about 60,000 years ago, there is evidence to show that July temperatures reached an average of 10° C (50°F) or more, while in January these figures dropped to a chilly −10° to −15°C (14° to 5°F). Not surprisingly, the hunters at Érd passed the winter elsewhere and spent only brief periods in spring and summer at the site. During the mild season, the surrounding landscape was not so desolate or forbidding as might be imagined, for among the 15,000 animal bones that can be identified, a wide range of nearly fifty species is present. However, despite the presence of the bones of all these animals in small quantities, the Mousterians did not practice a "catch-as-catch-can" policy, an indiscriminate plundering of every species that they encountered. Instead, for a brief period of no more than two or three months in early spring, they concentrated their attentions on either newly born or one-year-old cave bears. The corpses of the bear cubs, which far outnumber the other species represented at the site, were dragged back in one piece to camp, and certain joints were stored in the smaller of the two limestone basins. The same hunters appear to have reoccupied Érd in the summer, when they left behind an identical range of tools but based their hunting mainly on horses and to a lesser extent on asses, rhinos, and deer. Unlike the bear corpses, the horses were dismembered on the spot, and only the head and limbs were carried back to base.

It is possible, of course, that this selection of certain animals was less the work of man than of nature and simply reflected the creatures that were most abundant or vulnerable in the local landscape. However, the heavy concentration on bears of such a narrow age group does suggest a deliberate human policy. Moreover, at the nearby site of Tata, set in an almost identical environment and occupied at roughly the same period as Érd, people using quite different tools concentrated heavily on the hunting of young mammoths. The lesson seems to be clear: long before the appearance of fully modern man or of the art objects of the Upper Paleolithic, people were already picking and choosing from the animal resources that surrounded them and were developing specific hunting tactics that they used successfully for centuries.

The hunting of some animals and not others is perhaps not surprising; all carnivores are to some extent selective of their prey. The intensive hunting of one species is a trait that seems to stretch very far back in the human record. Sites with the consistent remains of young animals selected from just a few species go back at least as far as the Mindel glaciation, at least 300,000 years ago. However, a heavy dependence on one particular game animal is a feature that appears to spread and intensify in Europe with the appearance of more and more sites preserved during the last Ice Age.

What advantages could this specialization have brought? An intimate knowledge of the daily movements and behavior of one type of animal must have lessened the element of chance in hunting, and perhaps in some regions, this became a vital factor as climates became more severe and game more sparse. Equally, such a dependence exposed human communities to the peril of overexploitation, of pushing a particular herd beyond its limits and into extinction, facing the human group in turn with a sudden crisis of resources. The swiftness of this process should not be underestimated. Small Siberian family groups (of herders, not hunters) were totally dependent during the last century on reindeer, and once the numbers of these animals dwindled to a certain minimum of a few hundred animals, the herds were particularly liable to collapse under the impact of a whole range of possible dangers:

> Any inclemency of weather, disease, man or animal with respect to such a herd is likely to reduce it suddenly and catastrophically below a point to which it can maintain either the family or itself.[3]

It seems likely that the concentrated hunting of a single species would have exposed the hunters to greater risk from this sort of abrupt natural crisis.

An attempt to reconstruct the likelihood and impact of the failure of prehistoric reindeer herds on Ice Age hunting bands was the imaginative work of one London University lecturer, Nicholas David. His case study concerned the Perigordian culture, which became dominant in the Dordogne region during a comparatively mild interval of climate from about 26,000 B.C. At a slightly later period, beginning around 23,000 B.C., a distinct group of tools appears at several classic sites, notably the caves of Arcy-sur-Cure at its northernmost limits and Isturitz in the Pyrenees far to the south. The group is distinguished from the typical Perigordian mainly by its small, multiple-purpose blades with an angled cutting edge (known as Noailles burins).

The distinctive assemblages of these tools had persisted in the Dordogne for several centuries after 23,000 B.C., when it is clear that there occurred a definite decline in the numbers of sites occupied, suggesting a drop in population. After this temporary setback, the Noaillan flint-workers seem to have reasserted themselves, but with a slight change evident in their toolmaking preferences. They appear to have continued their activities until they "merged" with the Perigordian culture, after preserving something like 2,000 years of independence. The key problem is to account for the decline in population in the middle of this period, when no appreciable change is registered in studies of the comparatively mild climate. Did the Perigordians and the Noaillans suddenly quarrel over some intertribal dispute and resolve their grievances in a pitched battle somewhere along the banks of the Vézère?

David's close examination of the problem draws attention to the interesting fact that the economy of the Noaillan hunters in the central area of the Dordogne was different from its farthest outposts at Arcy and Isturitz. The animal bones recovered from excavations in the border regions demonstrate that the communities there exploited a broad range of quarry, including wild horses, red deer, and bovids. But in the valleys of the Vézère and the Dordogne, the Noaillan people relied almost exclusively on reindeer, which accounts for over 90 percent of the bones recovered from living sites.

This concentration on reindeer, David suggests, was a potential source of instability for the hunting groups, and is a possible clue to the mystery of their sudden depopulation. Drawing on comparisons with the recent caribou hunters of Canada, David suggests that the reindeer could have been intensively exploited beyond the point at which they could successfully replace their dwindling numbers. Once the population of the reindeer had fallen below a certain minimum, the fate of the entire herd was swift and certain, and with it, the dependent hunting group also was doomed. A comparable failure of reindeer herds in late nineteenth-century Canada was provoked by the overspecialized hunting of the Naskapi Indians with their newly acquired rifles, hundreds of whom starved to death as a consequence. Those Paleolithic people who were fortunate enough to have survived the catastrophe of the collapsed reindeer herd might possibly have combined together some time later to form new hunting bands, accounting for the slight differences observable in the tool kits of succeeding centuries. Meanwhile, the communities in the Pyrenees and in the Cure valley with their varied and relatively secure subsistence continued to flourish throughout this period without any apparent calamities.

David's attempt to portray an economic crisis of 25,000 years ago is highly original. Few prehistorians have ever tried to reconstruct a sequence of events by putting together the evidence of tools, animal bones, and climate in this vivid way. The Noaillans cease to be characterized by a tedious, static collection of cave sediments and flint tools, for in David's case study we imagine them as flesh and blood, faced with the possibility of starvation. While acknowledging

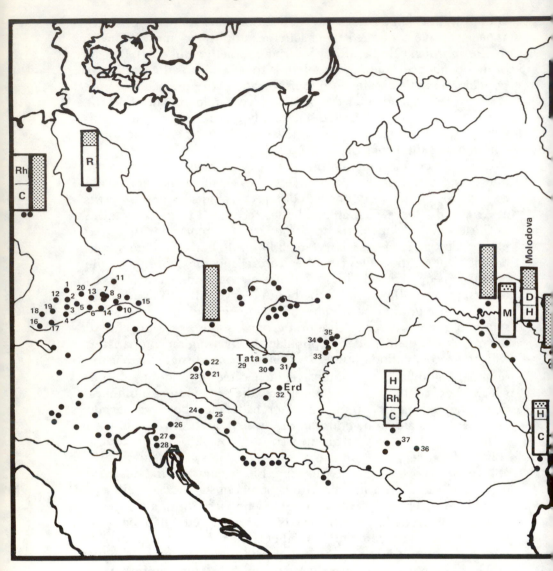

The quarry of hunters during the Mousterian period in central and eastern Europe. Each column represents the total animal remains at a particular site, divided into percentages of the main animals present. Sometimes the exact percentages were not properly recorded by the original excavator, and in such a case the column is shaded or else divided into equal segments.

The map suggests that specialized hunting was practiced at 17–18 Mousterian sites. The record could simply reflect which animals were most numerous or vulnerable in a particular environment, but heavy concentrations on a single animal are likely to be due to some degree of human choice.

This map is a much simplified summary of the available evidence. Not every Mousterian site is shown, but the black dots give some idea of the main centers of settlement in the early Ice Age. For further details, see Gábori 1976 pp. 197–206.

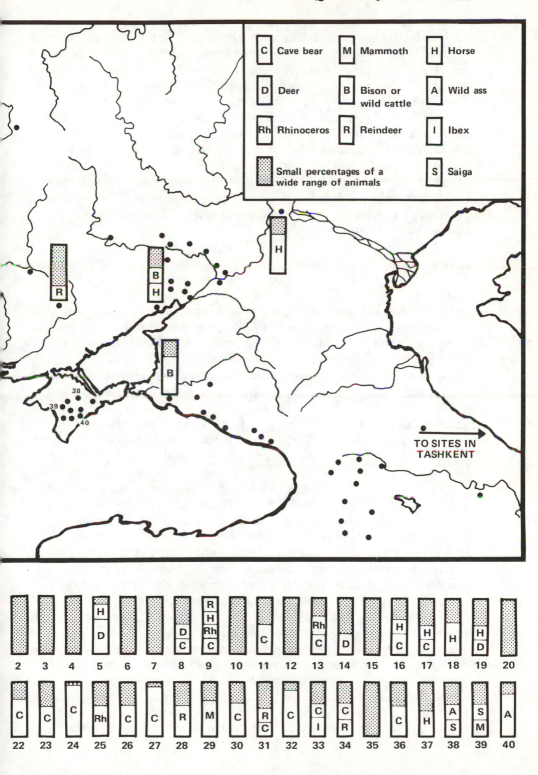

the originality of this work, however, we must also recognize several serious drawbacks, including a refusal by some French archaeologists to consider the Noaillan tools as representing an ethnic group distinct from the Perigordian. The study is based mainly on just three related Arctic hunting groups. The problems that these imaginary Noaillans encounter are specifically those suffered by Eskimo bands during the last century. As we have already seen, the inhospitable environment of the Arctic is not comparable to the Ice Age world with its richer vegetation and much broader range of game. Can we seriously imagine the Noaillans continuing their policy of intensive reindeer hunting and not falling back on other resources if they found themselves in imminent risk of starvation? Some of the unfortunate people who lived in the bleakest regions of Canada or Siberia had no such options.

The existence of "emergency" food sources in the Paleolithic period is scarcely to be doubted. At many sites in western Europe where prehistoric hunters concentrated on just one large game animal, there is often evidence of a broadening consumption of other types of food. If it were efficient to exploit big game by specializing in only one species, then surely the hunters would not inefficiently neglect to extract the most from vegetables, fish, birds, and perhaps even rodents or shellfish.

An instructive case is the recent careful study of animal bones from the coastal region of northern Spain, many of them excavated from the deposits littering the entrances to painted caves. Reindeer were present along this coastal strip during most periods (there are a few paintings and engravings of them on the cave walls), but their bones form an insignificant proportion of the debris dug up at the sites. Because of milder and more humid climates south of the Pyrenees, it is likely that reindeer were never present in the great herds that must have grazed on the banks of the Vézère and Dordogne. There, the Noaillans and Perigordians could concentrate almost exclusively on reindeer, while their contemporaries in Spanish Cantabria followed a much more varied hunting strategy. Only in the succeeding Solutrean period are there any signs that the Spanish hunters were starting to favor one type of game over another. From this stage onward, the bone records show a swing toward red deer hunting in the forested coastal valleys and in the increasing numbers of ibex sites in the rugged, upland Basque country.

Among the coastal communities lived one or several brilliant artists who lay on the floor of a cramped chamber at the cave of Altamira and painted vivid red-and-black bisons on the low ceiling. This chamber adjoined the main occupation space in the cave not far from the entrance. The people who lived under the porch of Altamira in the centuries around 15,000 B.C. were almost certainly

Top, the famous painted ceiling at Altamira, northern Spain, and, below, a view near the cave entrance (under trees on skyline in center of photograph). To the right, the ground drops down to the valley of the River Saja.

surrounded by a classic "mosaic" landscape. There was probably undulating heathland stretching down to the coast a few kilometers away, while in front of the cave porch, the ground dropped away to the valley of the river Saja, where herds of red deer, roe deer, and boar doubtless sheltered in winter among the tree-clad slopes. During this season, the hunters concentrated on red deer hunting but also fished down in the valley (a few fish vertebrae have survived in the bone collections) and gathered shellfish at the coast. Alongside numerous mollusk shells, a single seal bone, possibly from a stranded animal, was found in the cave. In summer, it is likely that the inhabitants moved to some campsite nearer the mountains to the south, where ibexes would have provided a likely target. While the quantities of red deer bones found at Altamira do indicate that specialist hunters lived there, the wide range of other animals represented in small numbers suggests that there would have been little risk of starvation even if the deer had been severely overhunted.

Indeed, fish bones are consistently present at other northern Spanish caves as soon as the swing to red deer and ibex begins in the Solutrean period. Later on, during the Magdalenian, thick deposits of shellfish appear, sometimes far inland (such remains at Castillo appear to have been brought about twenty-five or thirty kilometers from the prehistoric shoreline). While big game hunting became more finely tuned to one major species, trapping was not neglected. During the Magdalenian period, there is a dramatic increase in the numbers of small carnivore remains (eighteen foxes are represented at one Basque country site, eleven mustelids at another, and so on). This broadening of food resources must have helped to balance some of the fluctuations and uncertainties of an economy based largely on the habits of a single creature. The proof seems to lie in the increasing numbers of sites that are recorded toward the end of the Ice Age, which presumably reflect a growing human population.

A similar "explosion" in the number of occupation sites and decorated caves occurred in the final phases of the Magdalenian throughout southwest France. Unlike northern Spain, where specialized hunting seems to have developed slowly, the ancestors of the French Magdalenians had been relying heavily on reindeer for many thousands of years. What, then, could have encouraged this later, steady growth of settlement?

Simple economic factors may only be part of the answer, but obvious refinements in hunting weapons seem to point to increasing expertise in the chase and a widening range of available foods. The development of skilled fishing is usually said to date to these last few centuries of the Ice Age. This was when such items of equipment as harpoons, fish gorges made of bone and antler, and little fish-shaped bone fragments that may have been lures become common in the

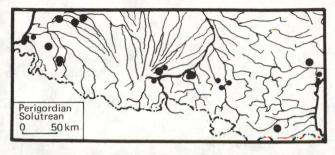

Population growth in the Pyrenees during the Magdalenian period. The top map shows sites of the preceding Perigordian and Solutrean cultures (the Solutrean sites are the big dots, in some cases located at earlier Perigordian sites). Despite the fact that the top map represents a longer time span, fewer sites are present.

archaeological layers. Objects carved with representations of fish, usually of the salmon family, belong mainly to this period, although the engravings appear on a wide range of material and not just on the surfaces of appropriate equipment like the harpoons. Finally, the large number of low-lying camps close to riverbanks dating from this time (such as the classic site of La Madeleine itself, a stone's throw from the river Dordogne) suggests that fishing made an important contribution to the diet.

The less conspicuous evidence of the fish bones themselves has been ignored until recently. Too often shoveled out of deposits unseen among masses of clay and rubble by earlier diggers, fish vertebrae now provide the modern expert with precise information about the species that were caught and, in some cases, the season of the year at which it took place. Thanks to this new research, we know that even the earliest inhabitants of Europe were eating a wide variety of fish, presumably caught by spearing or wooden devices. We know that a group of Mousterians in the south of France enjoyed feasts of trout, eel, and carp, although none of the surviving bone or stone work suggests the existence of any special fishing gear. The degree of care and precision devoted to the manufacture of bone or antler harpoon heads in the Magdalenian implies that the spearing of large fish, such as trout, salmon, and pike, had become a particularly desirable or prestigious activity. On the other hand, the previously unsuspected presence of more elusive creatures, such as roach, dace, eel, and burbot (revealed by their vertebrae), indicates that other devices—traps and nets, perhaps—were being used. Before

An Eskimo family stocking up for the winter, top, photographed in northern Canada in 1964. The char were caught in large numbers during their autumn run and then stored under stones for the winter. Below, a carefully observed engraving of a trout excavated at the Grotte des Deux-Avens, Vallon, southeast France, in 1969.

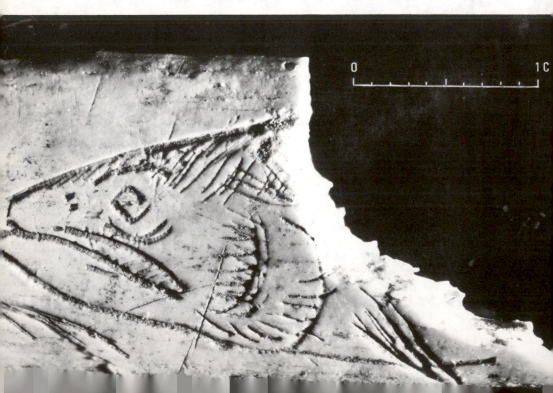

the coming of pollution and high dams, the major rivers of the southwest were famous throughout Europe for their salmon runs, and such abundant and reliable resources must have contributed to the security of Paleolithic life.

Just as fishing "took off" toward the end of the Ice Age, so, too, did fowling. Bird bones bearing the traces of cutting marks where flesh was removed only become widespread at Magdalenian sites, and they generally belong to small game birds, such as grouse and ptarmigan. Larger species were also hunted sometimes, as at the Abri Morin, where the claws of numerous great white owls had been removed, although we cannot say whether the owls were caught for their meat, plumage, or claws. The dramatic increase in Magdalenian bird remains is one possible argument for the invention of the bow at this time, although the existence of arrows is only directly proved by wooden shafts preserved at a north German site dating to about 9,000 B.C.

Fishing and fowling must have reduced the pressures of big game hunting; as the chase intensified, the existence of prehistoric communities seems to have become more secure rather than less. Given their skills and success, is it possible that the hunters sought to control the unpredictable aspects of animal behavior? Is it conceivable that such close relationships were forged between man and beast that active steps were taken to tame or herd the animals on which they relied?

6: The Legion of the Deer

The Problem of the "Bridles"

Although the cave bears that prowled the pine forests of Érd are not likely to have been tamed, there were other species that were far more amenable to human control. At the turn of the century an excavator of a prehistoric site (La Quina, in the Charente district of southwest France) noticed unusual traces of wear on the front incisor teeth of horses dug up from the Mousterian layers. These marks resembled those on modern teeth that result from the "tic" or nervous chewing habit of horses shut up in captivity, where boredom drives them to nibble incessantly at hard objects.

A few years later, in 1915, a lengthy study of some 16,000 modern horses was published, comparing the teeth of those kept tethered up with others roaming in relative freedom on the North American prairies. This study concluded that the nervous "tic" and its associated pattern of tooth wear is never present on animals free to wander at will. Could it be, then, that some Mousterian communities kept corrals of horses at certain times of the year, facilitating a control over animals normally undreamed of for this remote period or for many thousands of years afterward? It may be significant that the excavator of La Quina also found no less than seventy-six of the mysterious stone spheres in the Mousterian layers that, if they formed parts of bolas (as suggested earlier), could have been used to bring down and capture animals alive. In any case, the problem of the peculiar tooth wear brought to light sixty years ago deserves to be reexamined with modern techniques.

A more startling claim, first advanced in the 1870s and revived recently by a British scholar, Paul Bahn, is that we may be able to recognize actual animal bridles among Paleolithic bone objects. This may seem a very far-fetched idea indeed, considering that horses are conventionally thought to have been domesticated in Asia sometime after 3,000 B.C. The arguments concern a strange type of object known by the picturesque term "commanders' maces" *(bâtons de commandement)*, which nevertheless do not seem to have been items of military insignia. These objects take the form of pierced staves of reindeer horn or bone, with a hole smoothly worked all the way through the branching point of the antler, usually shaped to a T or Y outline. Batons without any form of surface decoration are known as early as the Aurignacian period from sites in the Dordogne, and they continued to be manufactured in increasing numbers right into

106

A *bâton de commandement* from La Madeleine in the Dordogne region, and *(right)* an undecorated example from the Russian Ukraine, shown here roughly at the same scale.

the final centuries of the Magdalenian. As in the case of so many other forms of decorated bone work, elaborate animal carvings do not appear regularly until the middle Magdalenian, when depictions of horses are the most common theme. During this period, prehistoric engravers concentrated more attention and accomplishment on the batons than on any other item of equipment apart from the spear-thrower, which suggests that by this time the curious rods had acquired a special significance. The Magdalenian carvers seem to have delighted in the difficulties of decorating the narrow, rounded surfaces with a fine tracery of complex and well-proportioned animal forms. Today, we can only grasp the total design when it is "decoded" in a single flat plane with the help of a cast of the entire rolled-out surface.

The quality of this carving and the fragility of some of the actual batons themselves make it difficult to believe that certain examples could ever have played a practical function. The idea of a sacred, symbolic wand may not, then, be improbable. During the seventeenth century, sorcerers among the Lapp reindeer herders used an identical type of object to beat a magic drum while they were in the midst of prophetic trances. Nevertheless, the signs of wear on some batons, especially around the inner edge of the hole, have inspired many ingenious, practical theories. One idea compares the prehistoric batons with identical objects worn by some recent Eskimo hunters as a kind of necktie under the throat. A more likely suggestion, one that is widely accepted by archaeologists, is that the batons were used as an aid in correcting the natural curvature of antler and ivory shafts intended for use as arrows and spears. According to one reconstruction, the shaft material was first softened over heat and steam; then one end of the shaft was wedged in the hole and the rest bound tightly under tension against the long arm of the baton until it became straight. However, the only researcher to study the traces of wear around the holes concluded that the marks resulted from the friction of a soft strap (so could they be handles for slings, this investigator asked?).

Yet another theory—the one that chiefly concerns us here—is that the batons formed the solid cheek pieces of leather or fiber harnesses that were slipped over the heads of horses or reindeer. In practical terms, such a use is quite convincing, for exactly similar antler pieces were traditionally employed in Sardinia for controlling horses through pressure exerted on the muzzle. Similarly, the Samoyeds of Siberia exercised some sort of control over the domesticated reindeer that pulled their sledges by tugging at the reins joined to simple head collars through the holes in pierced antler staves. Despite the feasibility of the bridle theory, there is no obvious reason why we should prefer it to any of the other ingenious ideas advanced to account for the batons; indeed, to accept it would mean adopting the unconventional view that horses or reindeer were domesticated by Paleolithic hunters.

Some fascinating and ambiguous evidence provides unexpected support for the bridle theory, however. Like most of the early antiquarians who had excavated in the Ice Age caves, Édouard Piette for many years regarded the horse and reindeer bones he recovered as simply representing the remains of wild animals brought down by traditional hunting methods. In 1889, however, he startled the pre-

Horse bridles of a traditional type from Sardinia with wooden parts superficially resembling the prehistoric batons.

historians assembled at an international congress by delivering an address on "The Question of Reindeer Domestication," in which he maintained that the high achievements of the Paleolithic engravers and toolmakers must be the product of a stable, sedentary society, based on an economy of domesticated animals. What had changed his mind? Piette explained that he had found

> several carvings at Mas d'Azil where the horses have a nose-band. The semi-domestication of these animals is thus well-established. Nothing proves that man harnessed them, nor made them pull loads, nor give milk, but why should they not have raised them in herds and have known how to lead them?[1]

Piette's views, stated modestly enough at the conference, became increasingly fervent as the years passed and as more of the curious "cut-out" carved horse heads came to light. The climax came in 1893, when an engraved bone piece was discovered in the cave of St. Michel d'Arudy, situated, like Piette's famous site at Le Mas d'Azil, in the foothills of the Pyrenees. This spirited carving shows a horse head oddly divided up into panels, with an abstract row of chevrons partly covering the lower cheek. This, for Piette, was incontrovertible evidence for the existence of head collars and the final proof of his domestication theory. His views, however, awoke bitter and at times unreasonable opposition among the leading prehistorians of the day. Even the youthful Abbé Breuil, who considered the problem at length, eventually changed his mind and decided against the reality of the bridles. He examined the entire range of engraved bone silhouettes and supported the idea that the curious lines and panels carved on the horse muzzles represented stylized impressions of their coat markings or muscular structure. Because the weight of contemporary archaeological opinion fell against his theories, Piette's views were largely ignored in the years following his death in 1906.

Reconsidering the problem of the bone horse heads today, it must be admitted that there is a considerable variety in these depictions, whatever their inspiration. Some bear no traces at all of the supposed bridles, while the clarity and detail with which the "cheek pieces" and "nose bands" appear on the other silhouettes vary considerably. One of the most delicate of the engravings was found lying on the surface of a cave floor near Arudy in 1975, only a few hundred meters from the site of the 1893 discovery. This new chance find has shaded lines and panels in exactly the same positions as on the St. Michel muzzle, although they are depicted much more faintly and certainly could pass as indications of the natural horse hair. Because of the peculiar, stiff style in which all the Pyrenean profiles are carved, a question mark will always hang over the exact significance of the patterns on their muzzles.

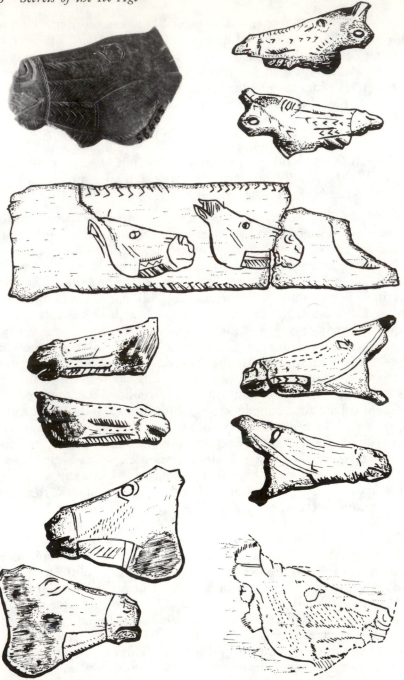

Engraved bone horse heads from the central Pyrenees, all bearing curious patterns that have been claimed to represent bridles or head collars. Not all at the same scale. Top photo: St. Michel d'Arudy. All the drawn examples are from Le Mas d'Azil, except *(top right)* Lortet and *(bottom right)* Espalungue, Arudy (the 1975 discovery).

An extraordinary engraved stone (published for the first time in 1966) suggests that the bridle theory should not be dismissed lightly. The stone came from the rock shelter of La Marche, hundreds of kilometers north of the Pyrenees, and shows a horse head engraved in a much more vigorous and realistic style than the Pyrenean examples. Unfortunately, the design, like many others at La Marche, is overlaid and obscured by a mass of fine scribbles. Among these, definitely carved later than the horse, are a series of lines that form the outline of a halter. These lines, whether or not they are the result of "doodling," are certainly not in the right place to represent the horse's face muscles (an explanation that Breuil and others favored to explain the patterns on the Pyrenean muzzles). So is the La Marche carving the strongest evidence yet of the existence of domesticated horses in the Upper Paleolithic? While the imagination may not take easily to the thought of mounted Magdalenians galloping through the Pyrenean foothills like Sioux or Cheyenne warriors, nevertheless some more limited usage of these animals, as Piette

Carved stone from La Marche, Vienne, western France, showing a horse head with numerous lines engraved over it, including some that closely resemble the outline of a halter.

originally envisaged so many years ago, may not be out of the question.

The Fickle Reindeer

One of the commonest charges laid against the early theories of semi-domesticated animals in the Ice Age was the absence of dogs, which would surely have been among the first beasts to be tamed and would have been invaluable for controlling large herds. The first dogs appear not long after the end of the Ice Age at hunting sites quite widely distributed across northern Europe, and it is likely that a long period of deepening relationships between hunters and wild dogs lay behind this development of domestication. The early stages of the process could well be imagined like the Aborigines' attachment to the wild dingo during the last century. Dingo pups were often captured by the aborigines and were sometimes suckled by women. When fully grown, they would be given names and fed regularly, or else treated rather indifferently. In fact, the adult dingoes were prone to desert the human group in favor of better feeding conditions in the wild and also do not seem to have been much use as hunting dogs (tracking wild dingoes and scaring them off their quarry at the last moment was far more effective). In short, the

Two Australian Aborigines of the Pintubi tribe in the Northern Territory, with domesticated dingoes.

A reindeer herd returning from its summer pasture, photographed in the Magaden region of Siberia.

raising of dingo pups seems to have started as a pastime more than as a practical measure. Something similar to this may have happened before the end of the Paleolithic. Since Piette's day, claims have been advanced for the appearance of dogs at several late Ice Age sites. Were these merely amusing pets, like the dingo pups, or could they have been used in practical hunting or even herding?

One of the very first facts to engage the attention of prehistorians was the frequency of reindeer bones at Ice Age sites. In particular regions, and for brief seasons of the year, reindeer provide a spectacularly abundant food supply. Consider the following description of the spring migration of caribou in the Canadian barrens, witnessed by a writer in the company of an Eskimo, both of whom watched awe-inspired as "the legion of the deer" passed close by them,

> each following upon the footsteps of the animals ahead. Here and there along the lines a yearling kept its place beside a mother who was swollen with the new fawn she carried. There were no bucks. All these animals were does, all pregnant, all driving inexorably towards the north and the flat plains where they would soon give birth, the surface of the bay for six miles east and west, had become one undulating mass of animals. The does gazed briefly and incuriously at us, and swung a few feet away and passed on to the north without altering their gait. Hours passed till the sun was on the horizon. . . . There was a sound of breathing and of moving that was like a rising wind. . . .[2]

Here, surely, was an unlimited stock of animal wealth for prehistoric people, a vast concentration of resources that they would not have failed to exploit. The reindeer, in the words of one prehistorian, was a "walking larder," an animal that could supply the total needs of subsistence if the undigested vegetable matter in the stomach and viscera were consumed with all the rest. Among the Eskimos, reindeer have provided furs and skins for clothing, tents, boots, and blankets, while strong thread could be made from the sinews. The fat provided oil for lighting, the bones grease for soup. A huge range of tools and hunting gear were manufactured from the bones and antlers. With the presence of these beasts in large numbers in Ice Age western Europe, the existence of prehistoric man would seem to have been untroubled and secure.

The windfall of the spring and autumn migrations, however, only lasted for brief periods. In a day or two, the hunters could kill dozens, even hundreds of reindeer, and then the animals would be gone, dispersed to their winter grazing in the forests or into smaller groups for the summer calving in cool pastures. Only during the autumn would their skins, enriched by the nutrients of summer grazing, be in an acceptable state for human workmanship.

Perhaps, like the most northerly Eskimo groups in recent times, Upper Paleolithic people regularly practiced meat storage by freezing or drying joints. Long-term "deep freezes" would certainly have been practical in regions with permanent ground frost; yet this must have excluded much of western Europe. (Even Érd, with its severe swings of climate for much of the early Ice Age, did not experience permafrost; for this, an average yearly temperature of $-2°C$ [28°F] or less is required). It should be possible, as was done at Érd, to study the bone remnants of such caches and to distinguish them from the normal debris of everyday living. Evidence for special meat-preserving areas is very rare at other sites, however. Either archaeologists have not looked hard enough at the evidence in the past, or else many Paleolithic communities had no need for stored food.

If we *do* accept the likelihood that dried or frozen meat aided the survival of hunters during the winter months, there is still the question of how secure a reliance on reindeer would have been in the longer term. All modern populations of reindeer undertake migrations of varying lengths, but the forces governing their movements are still inadequately understood. It seems likely that a number of factors, such as the avoidance of overgrazing in winter and of high temperatures and insects in summer, combine to drive the animals in their complex, ceaseless patterns of movement. Would the "legion of the deer" pass by with mathematical precision every year at the same time and place? In modern times, very few Eskimo communities relied exclusively on setting ambushes for caribou along their spring or autumn routes, and among those that did, a failure of the herd or an unexpected switch in its movements meant either whole-sale emigration or starvation. The movement of herds in the long term seems to have been extremely unpredictable, as one survey of Alaskan caribou in the 1930s concluded:

> the principal feature of caribou migration appears to be its uncertainty. One season the animals may come through in great numbers, massed in good-sized bands, a striking procession. . . . Another season their passing may be almost unnoticed as they dribble through in such small bands as to leave the impression that there are few caribou in the country.[3]

So we can imagine the Eskimo hunters waiting along last year's migration route, poised anxiously in canoes at the crossing points of rivers or hidden deep in the forest beside funnellike drives built of turf and brushwood. Yet most of them had other reliable resources, such as fishing or sealing, to fall back on if the caribou hunt failed.

Their precarious position contrasts strikingly with the impression to be gained of Paleolithic reindeer hunters. How can we account for the extraordinary persistence and abundance of reindeer bones

Caribou on migration through the Canadian Arctic.

in occupation sites in some regions, often spanning thousands of years? The evidence is most impressive in the "classic" cave region of southwest France around the Dordogne and Vézère valleys, for here, in over half the well-documented sites occupied by later Paleolithic hunters, the proportion of reindeer bones is greater than 90 percent and occasionally mounts to 98 or even 99 percent. These deposits represent century after century of successful human settlement, without obvious signs of long-term breaks, through a wide range of climatic conditions when other game animals, such as red deer or horse, must have been not only present but abundant. Why —and how—did the hunters choose to base their livelihood on such a seemingly fickle and unstable creature?

One answer could be that the behavior of modern caribou is uncharacteristic of the reindeer of the Ice Age. Just as the traditional subsistence and culture of the Eskimos were disrupted by the coming of European settlers, so, it is argued, might the movements of caribou have been grossly distorted by overhunting with rifles and a forced adaptation to the bleak tundra. The endless days and nights of the Arctic world, which certainly did not figure in the prehistoric environment of Europe, could have encouraged extreme migratory movements. Perhaps, after all, the Ice Age reindeer herds stayed relatively immobile during the year, only shifting from the grassy steppes into the forested "parkland" in a limited area from season to season. Indeed, perhaps a relatively stationary herd would have enabled a human group to occupy a single region or even one cave the entire year. Can we explain the splendor of cave art and the intricacy of flint tools by such a settled existence, exactly as Piette believed in the 1890s?

An expert of modern times, the distinguished French paleontologist Jean Bouchud, has endorsed this view in the course of his careful research. He proposed that the smaller "woodland" reindeer, which today lives mainly in forested environments and undertakes much less adventurous migrations, was present in the prehistoric deposits as well as the open-country "tundra" form. Clearly, there is no point in comparing the privations of modern caribou hunters with those of the Paleolithic if the latter were pursuing a more placid and predictable form of the same species. Furthermore, Bouchud studied the ages at which the prehistoric reindeer had been slaughtered, based on examinations of their antlers and teeth dug up at the cave sites and came to a remarkable conclusion. He proposed that the remains represented not just "windfall" hunting during a brief season of the year but that reindeer were killed all year round. The case for sedentary ancient hunters, procuring a steady and undemanding livelihood from their "walking larder," the woodland reindeer, seemed convincingly proved.

Subsequent studies by other experts have all tended to show the improbability of this theory. For example, recent work indicates the existence not of two distinct types of reindeer with different behavior and adapted to different habitats but only a single type that seems to have varied in stature considerably over the centuries. Even if we assume that this one type of beast made relatively slight seasonal movements, it has been clearly demonstrated that trying to exploit a herd from just one site the whole year through would have been impossible and, indeed, suicidal. It is a stark question of calories gained balanced against calories lost in perpetual movement. The hunters would have starved in their efforts to provision a home base even a modest distance from the farthest seasonal limits to which the reindeer wandered. It would have been much simpler and safer to pack up and leave the home base, following the herd for half the year. To reinforce this conclusion, it is now clear that the aging work that Bouchud performed on the antlers and teeth was founded on a number of misconceptions. In fact, much of these data could be taken to show seasonal, not continuous, occupation of the prehistoric sites.

Still we need to know how the prehistoric hunters of southwest France achieved that remarkable millennia-long record of dependence on the unpredictable reindeer. If the humans were migrating along with the animals, the pursuer and the quarry within constant sight and smell of each other, what sort of relationship would have developed? Could the virtual exclusion of other big game in the archaeological layers be less the result of phenomenally successful hunting tactics than the token of intimate herd control and management? As the centuries passed, it would surely have been obvious to the prehistoric hunters that they could not merely adapt to the creatures' instincts, but to some extent bend them to meet human requirements.

Hunters and Herders?

The history of man and reindeer in recent times is the record of a rich variety of relationships, some of them on a more mutual basis than others. At one extreme, the reindeer can be entirely socialized and subjected to human demands, dependent like a pet on human care and answering to its name when called, although usually maintaining a certain aloofness. Describing the fully domesticated reindeer kept by the Tungus of Siberia, one writer said:

> The reindeer is of a very mild and kind nature. It is attached to man and especially to those who use it kindly, speak to it, caress it, and generally pay attention to it.[4]

A Siberian Yakout on the back of a domesticated reindeer.

In this position, the reindeer will adapt itself to many uses, including giving milk (as early as 500 B.C., a drawing in a Chinese manuscript depicts this). A good reindeer is said to have pulled heavy sledges up to fifty miles a day. In addition to carrying loads, the Chukchi herders of Siberia used their sledge beasts for running down wolves and foxes, which gives some indication of the speeds they could attain. A more significant point is that domesticated reindeer were often used as decoys, sent in among wild herds and directed by the hunter's calls toward an ambush. Unfortunately, animals used in this way were often prone to desert their master and revert to their wild state. The larger the number of domesticated animals owned by the hunter or herder, the less likely he was to lose them.

A similar development among the Paleolithic hunters of France is by no means impossible: a number of tame reindeer, owned by a hunting band, might have been kept as draft animals or as a means of exercising an indirect control over the movements of free-roaming beasts. The northernmost people of Europe, the Nganasan hunters of the Taimyr Peninsula, adopted this practice in recent history. Unlike the Chukchi, the Nganasan never closely supervised their herds but simply followed the deer along their annual 700-mile migration trek. Yet tame reindeer found a wide variety of uses; for example, in summer, sleds were occasionally driven as much as 100 to 200 kilometers away from base on hunting expeditions. During the autumn rutting season, the antlers of tame males were used to ensnare those of wild ones. As the males engaged in combat, the antlers locked together, and a noose carried on the antlers of a tamed animal would trap its adversary, allowing a hunter to rush out of hiding and kill it. The Nganasan also used "bolas" made of knotted ropes and developed ingenious hunting nets and methods of driving.

Unlike sheep or goats, both horses and reindeer undergo no obvious physical changes when they become fully domesticated. We cannot examine a particular set of reindeer bones dug up from an ancient site and tell whether it was a placid pet answering to its own name or a creature of instinct living entirely in the wild. Yet one or two tantalizing hints exist. In the Magdalenian layers at the cave of Isturitz in the French Basque country, the leg bone of a reindeer was dug up showing traces of a serious fracture, accompanied by osteitis and running sores; yet the animal had not only survived this injury but, according to one estimate, had lived for another two years or so afterward. Could an animal with as serious an injury as this have evaded predators for so long in the wild? Or could this reindeer have been a tamed animal, cared for by the Pyrenean hunters at Isturitz?

In practical terms, the step from reindeer hunting to reindeer herding was a short one, while the leap in social values could be considerable. The more that hunters concentrated on the one resource, tracking it throughout the seasons, and the more they devised hunting drives or lured wild herds with the aid of tame beasts, the more intimate the control over the animals' movements would become. The changes in technique from hunting blinds to corrals, from herd following to herd supervising, might have been unnoticed over the centuries. On the other hand, the problems faced by all human groups that depend on reindeer herding are considerable, and their solution could take many forms, some of them with serious social consequences.

True pastoralism, which involves the intensive roundup of stock with the aid of corrals, was a comparatively recent development during the last century among the Lapps of Scandinavia. Previously, reindeer hunting was based on traditional, communally owned territories. However, the excessive hunting of wild reindeer stocks eventually grew to the point where joint ownership broke down, and

A print published in 1827 depicting a Lapp family setting out on the summer migration.

the individual's supervision of his own herd developed. The prestige of the Lapp no longer stemmed from his skill as a hunter or his ownership of a kill but became directly measured in the number of reindeer under his control. Wealth was slowly but surely transformed into a living asset, a commodity rather than a human quality. The communal values of a traditional hunting society broke up, and in their place came a more complicated economic and social system.

Yet in Finland, the pastoral solution was only temporary, itself carrying the seeds of instability and eventual breakdown. The herded reindeer were not, like the tame decoys, slavishly dependent on human whim, for the relationship was a mutual one. The reindeer supplied the herder with meat and skins. In return, the herder kept natural predators at bay and encouraged maximum breeding through the selection of males. The result was a population explosion, quickly building up enormous pressure on the grazing pastures; among the Skolt Lapps, between 1829 and 1909, reindeer herds are said to have increased thirtyfold. Eventually, despite huge losses during the Second World War, the herds grew so unmanageable that modern machines—the gas-driven snowmobile, light aircraft, helicopters, and walkie-talkies—were introduced during the 1960s to cope with roundups. The effect of these devices was to frighten and fragment the reindeer population, breaking up the mutual relationship of herder and animal, so that the entire system has now reverted to a kind of crude mass hunting, this time with high-powered and noisy weapons.

What are the lessons of Lapland reindeer exploitation, if any, for the prehistoric world? Obviously, some of the pressures on the modern pastoralists, such as the growth of a European market economy in Finland, were not present in Paleolithic times. On the other hand, it seems likely that intensive efforts to herd the reindeer would be accompanied by the kind of rapid population increases that are recorded in nineteenth-century Lapland. We can imagine that pressure on grazing land might have affected many of the traditional rights and relationships shared between local prehistoric hunting bands. Could such a major disruption of the social fabric have happened in Ice Age France without leaving some traces behind in the archaeological record?

The Chukchi of eastern Siberia are one people who coped with the problems of close reindeer herding for many centuries. To a limited extent, the reason for their success appears to have been more or less constant warfare against neighboring tribes. During the nineteenth century, their herds expanded eight- or tenfold, while the actual human population is scarcely likely to have even doubled. The forces behind this expansion probably have less to do with the natural multiplying of the Chukchis' own reindeer herds than with the forcible seizure of those belonging to the Lamut, the Koryak, and the other adjoining tribes.

In modern Siberia, reindeer herding offers a more secure existence than formerly. Pesticides check diseases, while helicopters and walkie-talkies assist with control of herds. Here Boris Ikevev of the Pole Star State Farm tries out his 40 kilometer-range walkie-talkie in 1970.

Chuckchi herdsmen, photographed by the Kolyma River in Siberia in 1917.

The Chukchi herders do not sound very much like the generally peaceable hunter-gatherers of either the Kalahari or the Australian deserts. Instead, Chukchi society was intensely individualistic, glorifying physical prowess and violence. This found expression in innumerable trials of strength, wrestling contests, religious rites presided over by ecstatic shamans, heroic war tales, feuding, and a high rate of murder and suicide. The social ideal was "the strong man," the rich herdsman who might increase his herd by various means, including marriage contracts and outright theft. His prestige was enhanced by "assistants," usually young hired hands with too few reindeer of their own to maintain an independent status; sometimes the assistants were war captives. After several years, the assistants frequently acquired sufficient reindeer of their own to break away from their service and to set themselves up as "strong men."

Some of the stresses of Chukchi society undoubtedly arose from the unpredictability of the reindeer, their staple resource. Herdsmen at both ends of the social scale, the most impoverished and the richest, were in particularly vulnerable positions. As noted earlier, once the numbers of a herd dropped to one hundred or so, the entire structure of the herd was likely to disintegrate under any outside pressure. On the other hand, the biggest herds were more difficult to control, more likely to overgraze the tundra, and more susceptible to epidemic diseases.

Indeed, although under favorable conditions the reindeer populations would "explode" as they did in Finland, the natural hazards faced by both man and animals in the Siberian wilderness were

Two faces of Chuckchi society: the warrior and the herdsman.

formidable. Annual plagues of flu and hoof disease could wipe out entire herds (one man lost three thousand reindeer in three days), while less frequent outbreaks of smallpox and measles hit human populations. Extremes of weather and the depradations of bears, wolves, and wolverines added to the losses. At the other end of the scale, the "strong man" with an unmanageably large herd had to resort to ritual sacrifices or to dividing the ownership of animals among relatives. Some social institutions had the effect of balancing out the extremes: most families were highly mobile, while fairs and ceremonies drew people from great distances. The popularity of long-range visiting meant the rapid exchange of information about relatives who might suddenly fall on hard times. The annual reindeer migrations themselves provided an opportunity for such far-reaching social contacts.

A successful long-term existence based on reindeer herding is not, therefore, an impossibility, but such an economy succeeds only at a considerable cost. The cost was paid, among the Chukchi, by periodic natural catastrophes that limited the tremendous breeding potential of the reindeer and by a self-centered set of values that helped men to combat their insecurities. Such a system could conceivably have arisen in the Paleolithic, but with all its disadvantages, there surely must have been some strong inducement—such as a lack of alternative, more reliable resources—to promote the growth of the herdsman's unstable way of life.

It is certainly doubtful whether or not we can recognize the practice of herding in the proportion of animal ages or sexes excavated

from an archaeological site. Some years ago, it was frequently assumed that hunters chose their prey at random, so that the age range
of beasts dug up at a hunters' campsite was expected to be just like
that of a natural animal population living and dying in the wild. By
contrast, it was supposed that pastoralists could be spotted easily
because they selected special groups for slaughter (usually young
males). However, because it is now clear almost as far back as the
earliest archaeological records that prehistoric hunters carefully selected their prey, the argument is useless. The efforts of hunters to
conserve the does might be just as conscientious and effective as
those of a herdsman. Both practices could leave behind identical
results in the bone remains at an archaeological site.

However, the mixture of "hard" evidence and the lessons of
modern reindeer exploitation do suggest certain conclusions. One is
that the seasonal following of herds throughout their migration cycle
seems the likeliest explanation for the overwhelming numbers of
reindeer bones at many sites in southwest France. The hunters'
efforts may well have been assisted by the taming of a few animals.
The close supervision of reindeer herds in historic times, however,
often seems like an extreme solution to the problems of existence in
extreme environments. As we have seen, close herding in the tundra
resulted in rapid animal breeding rates, pressure on land, sudden
catastrophic losses, starvation, and even warfare. These features *could*
conceivably have characterized the Upper Paleolithic, too. But is
there room for a compromise, a halfway point between hunting and
intensive herding, that might fit the evidence better?

The most interesting material has emerged from the comparison
of sites spread over wide distances. In two fascinating papers, a
Cambridge prehistorian, Derek Sturdy, showed how a number of
central European reindeer sites, dating to the final centuries of the
Ice Age and sometimes separated by as much as 500 kilometers,
could best be interpreted as summer or winter camps at either end
of long, seasonal migration routes. In one case, Sturdy connected
caves occupied in the summer months in the south German highlands with another group of open-air sites many hundreds of kilometers to the north on the coastal plain, which showed all the signs of
settlement during the winter months only. Sturdy suggested that the
two distinct groups of sites reflected the annual comings and goings
of the same bands of hunters, who were intent on keeping a close
watch on the migrant herds, if not constantly keeping up with them.
It is interesting that the bones at one of the winter lowland sites at
Stellmoor indicate a narrow range of ages and sexes that happens to
coincide neatly with the practices recommended by reindeer ranch-
ers in some parts of Scandinavia today.

A freshly established reindeer economy in western Greenland

provided Sturdy with a means of vividly reconstructing the seasonal round of the ancient German hunters. Every spring and autumn, the herds embarked on their great journeys, sweeping along the migration routes and covering perhaps thirty kilometers or so in a day. At this point, Sturdy suggests that the hunters abandoned their settlements and themselves became highly mobile, traveling light and sleeping in tents. Human control, if it were exercised at all, could be very indirect, as Sturdy himself observed while following a herd in western Greenland:

> The Greenland reindeer are moved by allowing the deer to see or smell a man in a given position, so that they will move away from the scent or sight in a desired direction.[5]

Further control over their movements was achieved by driving a herd toward natural obstacles, such as hill ranges. When slaughtering was necessary, the animals were driven into small valleys with bottleneck exits or entrances that could easily be blocked off, so isolating a part of the herd almost as effectively as a man-made corral. Many prehistoric sites in Germany command key positions at the necks of such confined valleys. Sturdy suggests that these valleys were exploited, just as in modern Greenland, to separate a small proportion of the herd when required.

These tactics of the Greenland herd-followers are very different from the intensive, cruder practices once carried out by some Laplanders and by the Chukchi. In Greenland, the more remote and discreet the human presence is, the better, say the herd-followers. If the animals are free to wander in as near a wild state as possible, this will result in a fatter and healthier animal population. At the same time, natural predators will take their toll, and possible disastrous population increases may be avoided.

We do not know if the Greenland system will work in the long run, for the economy was only created thirty years ago. Its success so far may help to explain what we encounter in the reindeer bone deposits of the Ice Age. The depth and continuity of the layers suggest that they could represent something more than a simple hunting strategy. On the other hand, if the Magdalenians had adopted the intensive herding methods of the Skolt Lapps or the Chukchi, involving close human contact and supervision, the result over the course of a few centuries would have been some major transformation of the cultural fabric, for which we have no obvious evidence. The likeliest answer, as shown by Sturdy's work, is a compromise between the two extremes: the discreet practice of herd-following, always one step behind the animals, which left them to their own devices as much as possible.

What evidence we have suggests that Paleolithic people were never the passive victims of their environments. There always was selection and a skillful manipulation of their surroundings, including a range of relationships with different animal species, each of varying depth and closeness. Neither did their economic choices dominate every aspect of existence; there were other factors, such as the acquiring of mates and the lessening of social tensions, that governed people's lives no less forcefully than the wanderings of reindeer. The question of these human relationships and indeed our notions of the entire character of prehistoric social life are examined in the following chapters.

Modern Eskimo design depicting a caribou ambush on a river.

7: Feast and Famine

The Original "Alternative Society"

In 1751, economist Anne-Robert-Jacques Turgot threw over his ecclesiastical career and took up the first of a succession of distinguished administrative posts in the French government. His decision seems to have coincided with the writing of his celebrated essay "On Universal History," in which notions of human progress and perfection were expounded clearly for the first time. Christian providence took second place to inevitable processes of social and economic advance, which were divided into clear stages. The first of these belonged to hunters modeled on early travelers' accounts of the Indians of the Americas:

> Without provisions and in the depths of the forests, men could devote themselves to nothing but obtaining their subsistence. . . . Families or small nations widely separated from one another, because each required a very large area to obtain its food: that was the state of hunters. They have no fixed dwelling-place at all, and move extremely easily from one spot to another. Difficulty in getting a living, a quarrel or the fear of an enemy are enough to separate families of hunters from the rest of their nation. So they move aimlessly wherever the hunt leads them. . . .[1]

Turgot's concern with economic progress meant that he placed special emphasis on the surplus created by agriculture. Indeed, for the philosophers of this self-confident era of developing capitalism, the hunting stage was almost a prehuman condition of unbridled savagery. Turgot's contemporary, Goguet, imagined wandering men "so ferocious, that many of them devoured each other."

This "Neolithic prejudice" against a hunting way of life became deeply rooted in Victorian values of social progress and in the developing classifications of tools and weapons into successive ages of Stone, Bronze, and Iron. Within such a framework, the concept of a clear division between hunting and farming seemed natural, the one a passive and parasitic dependence on nature, the other the essential condition for the birth of civilized, settled life. The persistence of this outlook is still extraordinarily strong. In 1960, the distinguished American archaeologist Robert Braidwood could write about the appearance of Near Eastern farming communities in terms similar to those of Turgot two centuries before:

127

The hunter as "noble savage": an engraving published in 1870, inspired by the early discoveries in the Dordogne. There is no evidence for the hafting of hand-axes as shown.

> Before it were some half a million years of savagery during which small wandering groups of people—living sometimes in caves and sometimes in the open—led an essentially "natural" catch-as-catch-can existence.[2]

So the image of the anarchic, nomadic hunter, his life fraught with insecurity and disorder, is a deeply entrenched one. Low levels of population among hunting peoples are seen as a reflection of scarcity and inefficiency. Lack of material possessions is visualized as grinding poverty, the outcome of a society hard pressed by lack of leisure time as well as constant need. Prejudices like these, abetted by blatant

Richard Lee interviews a !Kung hunter about a recent kill.

racial and religious bigotry, helped Europeans to drive modern "primitives" far from their favored hunting grounds and into the marginal wastelands of tundra and desert.

The inevitable reaction to centuries of bias and distortion came when anthropologists began to observe the behavior of hunters and gatherers in systematic detail. The most important single influence in creating a new climate of opinion was a study by Canadian anthropologist Richard Lee of the !Kung Bushmen of the Dobe region in Botswana. This area is located in the northwestern reaches of the Kalahari Desert, cut off from civilization by an arid plain that takes three days to cross on foot or two by donkey. Lee's first season of observations was carried out in 1964, during a summer of almost unparalleled drought, in a landscape normally subject to abrupt and unpredictable changes in the meager summer rainfall (swings of as much as 250 percent have been recorded). Despite these harsh circumstances, the !Kung diet was abundant, varied, and apparently more reliable than that of the pastoral Bantu, who shared the Dobe region.

The secret of their "luxurious" existence was an ample plant food supply comprising over one hundred edible species, of which the Bushmen were aware of about eighty but only exploited three as major food sources. Of these, by far the most common was a nut called the mongongo, which was so abundant that, as Lee observed, "millions of the nuts rotted on the ground each year for want of picking." Yet because the mongongo's hard shell preserved it against drought and rain for several months, the Bushmen were able to harvest it almost like a cereal crop, although by comparison it was far more nutritious (the average daily picking of mongongos yielded about the same amount of protein as a pound of beefsteak).

Nor were such substantial meals achieved at the cost of tremendous effort. Lee's figures showed that the average time spent hunting or gathering was about fifteen hours a week, spread very unevenly and with long intervals of leisure. A day's collecting by the women supplied a family for the next three days, which would then be devoted to chores around the camp fire (but these only lasted for one to three hours a day) and to entertainments of every kind. The men tended to hunt for longer hours but were equally sporadic in their efforts and were much less productive than the women. So the !Kung passed their lives in an uneven round of feasting, gossip, sleep, and dancing, while the demands of work seem trivial compared to those of an industrial economy. It is not surprising that some writers, discussing the Bushmen and other hunter-gatherers in a climate of intellectual restlessness and revolt during the late 1960s, turned to the Kalahari as an idyllic Garden of Eden, and characterized its inhabitants as "The Original Affluent Society."

That title came from Marshall Sahlins, anthropologist at the University of Chicago, who pulled together Lee's work and other recent hunter-gatherer studies into a stinging, strident attack on "the traditional dismal view of the hunters' fix." Sahlins' influential essay was as much a direct assault on the materialistic values of his own society as it was a reevaluation of conservative anthropological views. The combination of both, Sahlins suggested, was responsible for the "dismal" outlook on the status of the hunter-gatherer:

> Having equipped the hunter with bourgeois impulses and palaeolithic tools, we judge his situation hopeless in advance.[3]

Sahlins' incisive essay stood most of the traditional views on their heads. He drew attention to a handful of early accounts of the Aborigines, where sympathetic observers had in fact commented on the well-being and leisure afforded by their way of life. Low population levels, constant mobility, and the lack of material possessions were the necessary conditions assuring a standard of living unparalleled in any other phase of human history. The invention of farming did not free men from material constraints, for civilization imposed increasingly higher demands on labor and natural resources. If the hunter were liable to face periodic shortages, these were insignificant compared to the widespread starvation of the modern world:

> Now, in the time of the greatest technical power, starvation is an institution.[4]

Meanwhile, the hunter's lot is difficult for us to grasp precisely because it represents the inversion of so many of our own values: restraint in population growth, restraint in material demands and possessions, and restraint in the social status of the individual. The wealth of the Bushman should be judged not by the few weapons and ornaments he carries from camp to camp but by the equality of his social relations and the untroubled leisure in which he spends the majority of his waking hours. If we envy the life enjoyed by scattered Bushman tribes driven deep into the Kalahari, how much more envious should we be of Paleolithic hunters who, as Sahlins notes, exploited better environments and indulged in rich ceremonial and artistic activities undisturbed by the impact of an alien culture:

> Fragile cycles of ritual and exchange may have disappeared without trace, lost in the earliest stages of colonialism, when the intergroup relations they mediated were attacked and confounded. . . .[5]

Sahlins, then, like many other anthropologists, emphasizes that most of what we see in the society of modern "primitives" is probably only the ghost of its former self, a pale shadow of a Stone Age past that we can scarcely envisage in all its wealth and complexity.

Sahlins elevates the !Kung Bushmen and certain Aboriginal groups with comparable, secure patterns of existence to the status of a prototype, an ideal model around which we can base all our thoughts and judgments of the hunting way of life, past and present. Speaking of the Bushman's irregular work periods, Sahlins half-seriously refers to the !Kung as practicing "that characteristic Palaeolithic rhythm of a day or two on, a day or two off."[6] When discussed in such a spirit, the Stone Age hunters and the Bushmen together sound close to that "alternative society" that many Western youths dreamed of establishing at the time of Sahlins' essay, the late 1960s. In the present climate of more cynical values, it is perhaps natural to wonder how representative the Bushman way of life really is, and if "The Original Affluent Society" was so untarnished an ideal. This is not to deny the justice of most of Sahlins' observations; yet if the values and virtues of the Stone Age are hard for a modern mentality to appreciate, so, too, may certain drawbacks of the hunting life be less than obvious.

The first, basic question is whether or not Richard Lee's meticulous work among the Dobe !Kung, around which so much discussion has revolved, was a true reflection of a traditional pattern of life. Recently, it has been suggested that the extraordinary abundance enjoyed by the !Kung was due less to nature than to the activities of the South-West African authorities just across the border. Despite police attempts to patrol the border, !Kung from the Dobe region of Botswana (then Bechuanaland) easily crossed over to hunt in the adjacent areas occupied by the Nyae Nyae !Kung. By 1964—the date of Lee's observations in Dobe—eight hundred of the Nyae Nyae had abandoned their traditional hunting grounds in favor of resettlement and agricultural instruction at a government station in South-West Africa. It is hard to believe that the disappearance of 800 people from a landscape shared by the Dobe !Kung would not have affected the ease and pressures of their livelihood in some fundamental way.

On the other hand, Lee's initial survey has been extended by additional seasons of work during 1967–73 and enriched by the studies of over a dozen experts in medicine, genetics, folklore, and archaeology. Their efforts were a desperate attempt to record the fragile pattern of !Kung life as it was steadily eroded by pressures from pastoralists and from the official takeover of their traditional hunting grounds for freehold tenure. One result of this new research is that the prosperity of Dobe is now clearly seen to be greater than other regions of the Kalahari. Elsewhere, the phenomenally nutritious mongongo nut was comparatively rare. The effect of this on the ≠Kade group was to encourage a broader diet than existed among the Dobe, based on thirteen major plant species instead of just three, and also involving travel over much greater distances. All this called for longer hunting and gathering hours than among the Dobe !Kung, in fact amounting to nearly double their work effort (although at an average of four and a half hours a day, it may still seem luxurious to us).

A more interesting fact is that, despite their superb craftsmanship and music, their storytelling and dancing, the !Kung do not sit around in blissful contentment, oblivious of future cares and uncer-

A group of Dobe !Kung: Does the apparent security of their hunting life bring contentment?

tainties. Surprisingly, they are reported as constantly expressing anxiety about food and its fair distribution, while their thin bodies have been noted by all observers. Indeed, the businessman's afflictions of coronaries and high blood pressure are nonexistent in the desert, but it has been demonstrated that most of the !Kung are in a state of mild undernourishment (although *not* malnutrition) from the time they are weaned onwards. Researchers say that this deficiency arises from a shortage of calories during just a few dry months in the spring season, when the diet is composed mostly of roots and bulbs.

So the fact that the !Kung are surrounded by extravagantly abundant resources in some regions and at some times of the year is fairly unimportant. What really influences the hunters' well-being, outlook, and mobility are the lean periods of scarcity, not the seasons of excess. From this point of view, the bestowing of titles like "The Original Affluent Society" on all hunting and gathering peoples is meaningless. Abundance does not always guarantee security and health in the long run, as we have already seen in the case of reindeer hunting and herding. Thinking that all hunters lead idyllic lives may distract us from considering the unique way in which each group responds to temporary bouts of feast and famine. Indeed, some anthropologists are forthright in their rejection of the idealistic view and express doubts about taking the !Kung to represent *the* classic example of the hunting and gathering life. Discussing cases of infanticide among Canadian Eskimos, two scholars recently said:

> There can be no question of the rigours of Polar life, and the current, slick rendition of all hunter-gatherer groups as the original "affluent society" bypasses the innumerable, tragic sagas of Arctic privations.[7]

There is little point in studying a people like the !Kung if we are to conclude simply that, by and large, they are well off and that Ice Age hunters were probably even more prosperous. Instead, we should be examining a variety of recent hunter-gatherer groups to see exactly *how* they have managed to cope with the basic problems of existence in uncertain surroundings. Of all these problems, one of the most interesting is that of the long-term control of population.

Birth Control in Prehistory

"Affluence" could be assured for a hunting group at what seems to us a fearful cost, by keeping the number of mouths to be fed at a strict minimum. The practices of infanticide, abortion, and feuding have been recorded in varying degrees among recent hunter-gatherers and so, too, have casual, careless attitudes to infirmity or disease.

British anthropologist James Woodburn reports this as a feature of the Hadza of Tanzania, who are a people

> strikingly uncommitted to each other; what happens to other individual Hadza, even close relatives, does not really matter very much. People are often very affectionate to each other, but the affection is generally not accompanied by much sense of responsibility. If someone becomes ill, he is likely to be tended only so long as this is convenient. . . .[8]

Here, perhaps, is a repellent aspect of the hunting idyll, an unpleasant side to their values that cannot be ignored. Do the very qualities that seem so remarkable to us—for example, the absence of authority grounded in the private ownership of land or goods—in fact breed social apathy and neglect? High death rates among infants and the elderly might well be exacerbated by such casual attitudes, helping to assure a steady existence for the stronger members of the group.

How we can measure the sense of social responsibility that existed among prehistoric hunters is an awkward question. The deliberate burial of extremely young Neanderthal children (including a probable fetus at La Ferrassie in the Dordogne) *could* be seen as a sign of sensitive responses to the perils of infancy. Rather more positive evidence for a degree of social welfare emerged from excavations at the Shanidar Cave in the Zagros Mountains of Iraq. Of the nine Neanderthal burials at this site, one was of an arthritic man with facial injuries, including a blind left eye. He also had suffered a withered right arm and shoulder from the time of his birth, which was probably amputated in adulthood just above the elbow. The considerable wear on his teeth showed how he had probably used them to compensate for his deformed and useless arm. Yet when he died, at around 44,000 B.C. (accidentally struck down by a boulder tumbling from the roof of the cave, like several other individuals), this man was over forty years old and had obviously won a place for himself in the community. After the rockfall, his companions placed stones over his body, and the animal bones discovered lying on these rocks may have been part of a funeral feast.

Beyond the tumbling roof blocks, life at Shanidar was not without its risks, however. Another Neanderthal accidentally struck down inside the cave porch also bore traces of a stab wound on one rib, an injury that had partially healed during the week or so that elapsed before the fatal rockfall. Another violent end overtook at least one other Near Eastern hunter, an inhabitant of Skhūl Cave at Mount Carmel in Palestine, whose pelvis was perforated by a probable spear blow. A vertebra from a child's body dug up at the Grimaldi caves on the French Riviera was discovered to have a flint projectile lodged in it, and no doubt this, too, was a fatal wound. Usually, however, enthusiasts have eagerly attributed all sorts of damage, some of it incidental and modern, to lurid, melodramatic events in the distant past. It is still possible to read accounts of prehistoric "murder cases" that have been firmly disproved by the work of recent experts. The "wound marks" on these bones usually turn out to be damage inflicted by the crushing forces of burial beneath the overlying layers of earth and stone. And even where violent injuries are indisputably present, there is of course no way of telling whether these represent deliberate acts of violence or accidents sustained in the course of hunting and gathering.

If murder *were* involved, the chances are that these incidents arose from personal disputes and feuding, rather than from full-scale warfare between opposed factions. True warfare, whenever the anthropologist encounters it, generally appears as a feature not of dispersed and mobile hunting communities but of more settled societies that are usually practicing some kind of gardening, sowing, or herding—that is, with something to lose or gain in terms of resources or land. Of course, the primitive consciousness is not always directed toward such logical ends as acquiring material possessions; the reasons for aggression are often lost in "irrational" cycles of ritual belief and action. But the fluid territorial notions of the hunter and his few, simple possessions must contribute to limiting the scale of the hostilities. Among the Australian Aborigines, disputes do arise over women or trespassers on special hunting grounds or sacred sites, but any conflict inevitably involves kinfolk, drawn into the dispute from widely scattered areas and different bands. They are as likely to try to calm their aggrieved relative as to support his case, and on the whole, the disagreements remain on a small scale, sometimes resolved by formal duels or specially appointed "juries." Family might be set against family, perhaps, but rarely band against band.

Prehistoric hunting groups may, then, have been characterized by certain "antisocial values"—ranging from neglect of the sick to sporadic outbreaks of interfamily feuding—but it is hard to believe that such factors would have a steady, long-term effect in checking human populations.

Whether or not direct controls over the number of births were always necessary is a vital question. Scholarly opinion is sharply divided on this issue, some affirming that infanticide and abortion have always been essential to balance the numbers of primitive populations with their food supply. Others suggest that many such communities are in extreme, "emergency" situations, dislocated by European interference from their traditional pattern of life where birth controls were rarely necessary.

A difficulty of both points of view is that of obtaining the relevant information among modern "primitives." For an outsider, it may be hard to distinguish between "ordinary" infant mortality, including miscarriages, and deliberate practices, such as abortion. Understandably, women may be very reluctant to talk openly about such controls. Yet where peculiarly off-balance sex ratios exist in a census of a hunting group, the anthropologist has reason to suspect that infanticide is being practiced even if no direct evidence is available.

As systematic studies of recent hunters began to emerge during the 1960s, certain remarkable facts connected with population struck the experts. One was the scarcity of hunters on the ground; in most regions of the world, a density of more than one person per square mile was rarely reported, even in areas with rich and abundant resources that could undoubtedly have supported far greater numbers.

To understand this, it is necessary to remember that European scientists intrude on an alien culture, stay for a brief period, and then depart. To them, such behavior as the !Kung's anxiety over food despite millions of mongongo nuts rotting on the ground may seem quite unreasonable. But the human group is not the product of one or two seasons, but of generations of accumulating folk-memory, custom, and superstition. Once again, it is scarcity and famine that are most likely to influence belief and action. What community would willingly tolerate conditions in which plenty was assured for all during nine years out of ten, if in the tenth year starvation deci-

A depiction of an Australian Aborigine duel as witnessed by the explorer Carl Lumholtz in the 1880's.

mated many families? In other words, over many years we might expect a hunting group to find some way to maintain and adjust its numbers at a low level, to respect these fluctuations in their natural surroundings. Their low numbers would offer some margin of safety to insure them against the worst times.

In the 1950s, one anthropologist who worked extensively in Australia, Joseph Birdsell, attempted to demonstrate such ideas with mathematical precision. He took over one hundred Aboriginal tribal territories and showed that their size was related very closely to perhaps the key fluctuating factor in the desert landscape: the frequency of rainfall. Where the average rainfall was most scarce and meager, so, too, was the human population.

Birdsell believed that the Aborigines had found ways of shaping their societies to fit their surroundings in a state of perfect balance, which extended back for centuries if not for thousands of years. Among these ways, Birdsell considered that the regular killing of female infants, which was widespread among recent Aboriginal groups, was perhaps the most significant factor in checking the growth of population at a family level. Since most hunting groups were highly mobile, a mother could effectively carry and nurse only one child at a time, so a minimum of three or four years between births was essential. These arguments led Birdsell and many others to assume that infanticide was an essential part of the hunting way of life, past and present. This was the principal mechanism that allowed hunters to match the size of the bands so successfully to the unpredictable forces of nature; one recent researcher refers to "the finely-tuned state of hunter-gatherers."[9]

Some modern hunters *did* view infanticide as a practical measure with the cold-blooded detachment of an academic concerned with population figures. Several Australian Aboriginal groups are recorded as practicing infanticide because of direct anxiety over food shortages or a wish not to impair the economic effectiveness of the woman as a food gatherer. Yet conscious motives are not always paramount in the primitive mind. For example, female infanticide was sufficiently widespread among the Eskimos to provoke wife stealing and feuding, and yet there is no agreement among the experts over the exact reasons for its existence. (One says that the Eskimos did not wish to support unproductive members of society; another, that there was a concern to even out the sexes because of the number of men killed in hunting accidents.) But the Eskimos themselves were not driven by rational thoughts, according to the one detailed record of such practices that survives. Anthropologist Knud Rassmussen, who worked among the Netsilik during the 1920s, when female babies were being killed at the incredible rate of 80 percent, explained:

An Eskimo grandmother and child from Chesterfield Inlet, northern Canada.

> The Netsilik never think of reasoning with themselves but simply react with one event or another which may force itself upon their notice.[10]

And despite the high birthrate among the Netsilik, Rassmussen thought that the Eskimo communities would soon be extinct if the extermination of female infants continued on such a scale.

Rassmussen's prediction has been amply confirmed in an elaborate model of a hunting population constructed in 1974 by means of a computer. The object of this study was to see what would happen to a group of hunter-gatherers if it practiced infanticide regularly over a period of several centuries—a time span much greater than the short-term vision of the anthropologist. The results make surprising reading for those who have assumed that infanticide is a reliable method of trimming population growth. Even assuming fertility rates higher than those actually recorded among the Eskimo, the regular practice of infanticide above a remarkably low threshold of 8 percent leads inexorably to total extinction in the course of four or five centuries. In other words, the killing of female babies as a long-term device to limit population is not merely effective; it is genocidal.

The authors of this original study accept that acts of infanticide and abortion are widespread among recent hunters and gatherers, but they suggest that these are actually short-lived patterns of behavior, responses to particular crises that are aggravated by the pressures of European settlement. Once again, we find extreme solutions to everyday problems taking place mainly in the extreme environments of the Arctic and the Australian deserts. In undisturbed Ice Age Europe, the stresses and strains on hunting groups may have been quite different, and it could be that infanticide was widely practiced only in real emergencies or to eliminate babies with obvious abnormalities.

The demand for a mother to be constantly mobile, to have only one child at a time to carry and nurse, is not necessarily a crucial factor. It has been shown that in an average hunting band there would probably always be more carriers available than babies— assuming that relatives would aid in transporting children, which is in fact a regular custom even among nonhuman primates.

Would mothers and fathers have always needed to exercise a conscious policy toward the spacing of births? A fascinating aspect of the physiology of some recent hunter-gatherers is that, even where infanticide, abortion, or abstention is uncommon, phenomenally long spacings of three years between births have been recorded. The classic case, once again, is the !Kung, whose women complain that God is stingy with children and that they would like more. Among the !Kung, there are no regular inducements to eliminate children, and infanticide is only practiced in the cases of abnormal births, one of a pair of twins, or of elderly mothers who feel they can no longer cope.

The reasons for the low productivity of women among the !Kung (an average of only five births for each mother) are not obvious. Prolonged breast-feeding, for example, is only likely to inhibit ovulation for at most a year and a half. Recent medical studies of the !Kung have drawn attention to the thinness of their physiques and to the possibility that fat levels are directly related to fertility. Extra calories are required for the immediate demands of early nursing and carrying, and these, it seems, the !Kung mothers do not get. Under these circumstances, low reserves of fat appear to hinder ovulation. As soon as a child is weaned, the energy requirements decrease, and fat levels rise to a point where ovulation can take place. Is it possible that periodic food shortages or the energetic life of prehistoric hunter-gatherers would have trimmed their bodies to the lean, spare outlines of the Bushman? In that case, the mechanism of low fat levels could have operated to save mothers from excessive child-rearing or the regular practice of infanticide.

It is always tempting for an anthropologist to take present-day observations of a people like the aborigines and assume that they preserve a record of a traditional way of life that stretches back for generations. Elements of ancient customs and behavior obviously do persist in the face of colonial change. What is really at issue is the idea that over the years simple communities have reached a point of steady, stable balance with their surroundings. This is the way in which Birdsell and others have pictured the aboriginal way of life—"finely tuned" to the landscape and adjusted by human control through the practice of infanticide—and this pattern has been projected far back into prehistory. But this could be a static and unimaginative way of looking at the remote past, a way that may obscure some of the very questions we are trying to ask. Populations may have fluctuated wildly over the centuries for reasons that are not now obvious in the modern setting. We do not really know how important the "natural" forces of disease and infant mortality were for limiting population compared to human controls or how these forces may have changed over time. The prehistoric record must surely contain cases of hunting populations that failed and became extinct as well as those in successful, balanced adjustment.

The Case of the Missing Fingers

It is impossible to form an accurate picture of the risks of mortality and disease in prehistory. The small number of skeletal remains at our disposal do not in any sense represent one population but are separate finds scattered throughout Europe. Even if we had access to many bones from one region, it would be impossible to quote a mortality figure without knowing if the population as a whole was expanding or remaining steady, and this, as we have seen, is a debatable question. All that can be done is to look at the lists of bone remains and form a general impression of the risks of being alive 10,000 or more years ago.

The most striking fact is that, in the case of both Neanderthal and fully modern skeletons, the highest number of remains are those of young children, and this gains significance when it is remembered that fragile infants' bones are less likely to be preserved than adult ones. Once the perils of childhood had been passed, it is clear that most adults were extremely lucky if they survived into their forties. More male skeletons are present in these older age ranges than female, which is an interesting reversal of the present trend in Western society. However, even in a modern population, men will have more chances of survival into middle age than women if the general health hazards are high. So the unsatisfactory information available does suggest a high rate of infant mortality and a short average life expectancy, in many cases falling, perhaps, in the early twenties.

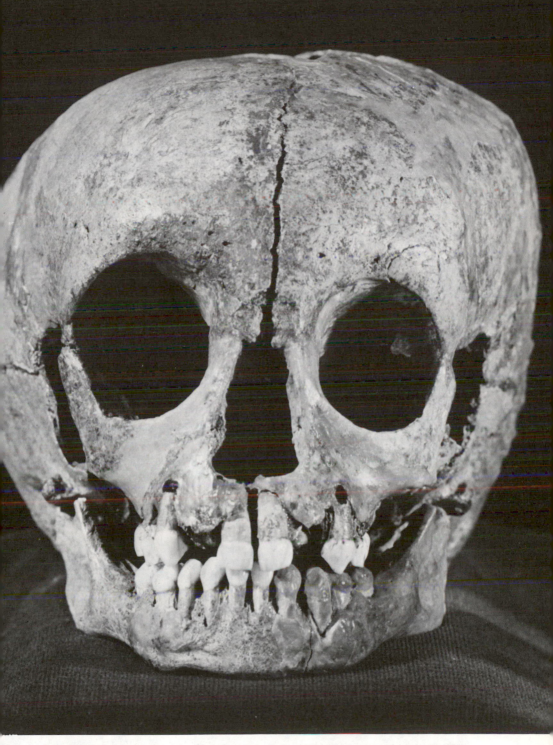

A token of the high risk of infant mortality in prehistory: the skull of a two-year-old child who died about 45,000 years ago at Pech de l'Azé in the Dordogne.

Even if Magdalenians had more opportunity than we do to dine regularly off grouse and salmon, the individual's chances of enjoying this "affluent society" were nevertheless extremely brief.

What were the main threats to life and limb? The danger of accidents in the course of an active hunting and gathering life should not be exaggerated. It is rather remarkable that so far no fracture of the arms or legs has been detected on any skeletal remains found in France. On the other hand, arthritic infections have left their marks on enough skeletons besides that of the "old man" of La Chapelle-aux-Saints to suggest that it was a common ailment. Contrary to the persistent popular myth that associates arthritis with cold and damp conditions, these infections are actually found in all climates and regions of the world. Its frequency among prehistoric Europeans was not necessarily a consequence of settlement in dank and dripping cave porches.

Ailments that result from a poorly balanced diet, such as rickets, are not evident among the ancient skeletons. This is not surprising because the prevalence of deficiency diseases in early industrial societies is generally blamed on lack of sunlight and inadequate meat consumption; thus, even with their unbalanced, meat-dependent diet, the Eskimos of the extreme north were nevertheless nutritionally sound. On the other hand, a rough diet was certainly connected with some of the dental problems suffered by prehistoric people, such as the severe tooth wear particularly evident on some Neanderthal teeth. Cavities have afflicted man since the earliest stages of human evolution, although they only became widespread after the development of refined flours and sugars during the early Middle Ages. A few cavities have been noted in Ice Age teeth, but dental complaints, such as abscesses, pyorrhea, and other infections (which could have fatal complications), seem to have been more serious problems.

The bones do not reveal any diseases that cannot be diagnosed by modern science, and certain maladies, including syphilis, appear to be absent in prehistoric times. However, many illnesses that can be halted swiftly and surely by modern treatment must have been greatly feared. One of the earliest of such cases is that of a nine-year-old child found at the Lazaret Cave near Nice, who probably died of meningitis toward the end of the Riss glaciation, about 120,000 years ago. Another concerns the original "old man" of Cro-Magnon, among the first prehistoric skeletons of fully modern type ever discovered, who has peculiar eroded patches on his bones, including a saucer-shaped depression on his right forehead. This used to be attributed, rather improbably, to damage by water dripping on the skull inside the porch of the shelter. All these traces have recently been related to a fungal infection called actinomycosis, which at-

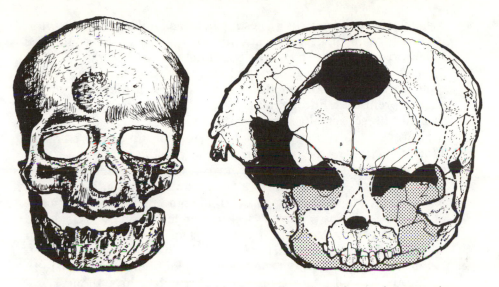

The Cro-Magnon skull and *(right)* the Rochereil child, not drawn at the same scale.

tacked the fifty-year-old man of Cro-Magnon chiefly in his jaw area, but is known to have fatal consequences in modern times if it becomes localized in the intestines. One can readily imagine the alarm and superstition that must have attended the onset of such diseases. Perhaps intricate herbal remedies were tried; the Neanderthal "flower burial" at Shanidar certainly indicates that a symbolic value was attached to plants at an early period.

Because it is impossible to form a reliable picture of the general health of prehistoric populations, the bizarre and exceptional ailments command attention. One of the odd cases concerns a child's skull that was found isolated from any other bone remains in Magdalenian layers at the cave of Rochereil in the Dordogne. This skull had originally been rested on two stones, but was badly shattered by the pressure of the overlying deposits. When it was painstakingly reconstructed in the laboratory, it proved to be that of a child between two and a half and three years old with an extremely round-headed appearance when seen from the front. The excessive width of the skull was almost certainly due to the condition known as hydrocephaly, or "water on the brain," which often creates psychological disturbances in the patient and, if untreated, leads to death at a very early age.

The most interesting fact about the Rochereil skull is that the upper part of the forehead bears a neat round hole about four and a half centimeters in diameter, where a circle of bone had been cut away with great care. We can imagine the anxiety created by this

Magdalenian child with its grotesque head and unbalanced behavior, leading to the final, desperate attempt to "operate" on the deranged brain. So far, no other skulls deliberately treated in this way (or "trepanned") are known from the Paleolithic period, although many quite normally proportioned skulls with holes in them are known later in prehistoric Europe. Sometimes fresh growths of bone show that the subject actually survived the procedure.

However, this is not the end of the story of the Rochereil child. When the edges of the hole were carefully studied, it became obvious that it must have been cut out *from the inside.* In other words, the roundel of bone was *not* cut away during an operation intended to relieve the child's suffering. Instead, at some time after the child's death, the skull was removed, and the circle of bone was detached by careful cutting from the inside; perhaps this disk served as a talisman to ward off the recurrence of this horrible affliction.

But the strangest ancient medical problem of all concerns hands, not heads. In well over twenty caves distributed throughout southwest France and northern Spain, there are dozens of painted hand impressions. These were created by pressing a real hand against the wall and applying color around the edges: the result was a "negative" print like those that children still produce. In one or two small caves, these stenciled hands form the main works of art, but usually, they are relatively inconspicuous, and it is animal compositions that chiefly catch the eye of the casual visitor to the decorated caves.

Several close studies of these hand impressions have helped to explain the techniques most commonly used in their execution. In most cases, a red or black water paint has been sprayed on, resulting in an even halo of colored droplets of various sizes around the handprint. Because the red pigment (hematite or ocher) is entirely

An intact hand outlined in black at the cave of Pech-Merle in the Lot region of southwest France.

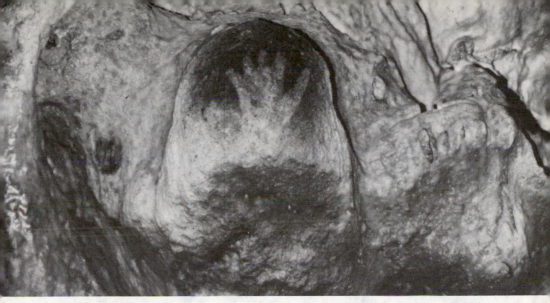

A mutilated hand outlined in black at the cave of Gargas in the Pyrenees. Are the fingers absent because of disease, a gruesome ritual, or "cheating" —folding down the fingers—by the artists?

soluble in water and the black (manganese dioxide) is not, the rock soaks up the red coloring more easily. As a result the red hands are usually more fuzzy and less clear than the black ones. It is uncertain exactly how the color was applied; it is usually supposed that the artist took a big mouthful of paint and blew it out between pursed lips to produce a fine mist of color. In experiments supervised by the author, the effect closest to the original hand impressions was obtained by blowing down an elementary spraying device made from a perforated reed.

Here, we might conclude, was a quaint pastime that we might attribute to idle amusement or to magical or religious rituals. In three caves, however, the bizarre nature of the hand outlines lifts the entire problem to a complicated and mysterious level. One of the three caves is called Maltravieso in northern Spain and features about thirty negative prints of hands. Only about a dozen are clearly visible, and of these, only one hand actually possesses all the joints of the fingers and thumb intact. On the other hands, there are two cases of missing thumbs and joints of the little finger. Elsewhere, in the caves of Tibiran and Gargas situated on opposite sides of the same conical hill in the French Pyrenees, more of the extraordinary mutilated hands appear.

Gargas has the most astonishing collection of these battered handprints. It is not a prepossessing cave; there are broad and rather low-ceilinged chambers, but the limestone is dull, and there are few glistening stalactites to divert the tourist. Yet in places, the sloping walls are covered with clouds of red-and-black hands, beginning close to the floor and mounting, halo after halo, almost out of conve-

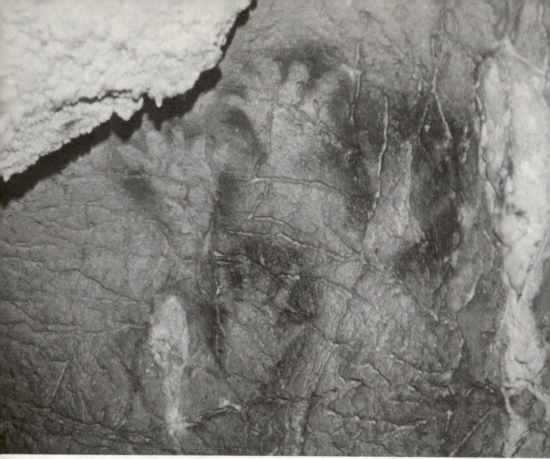

Mutilated hand impressions in the main chamber at Gargas.

nient reach. Unlike Maltravieso, there are no thumbs missing at Gargas, but the catalogue of incomplete finger joints is long and varied. Recent censuses of the hands have resulted in conflicting figures because some are faded and difficult to see. However, there are well over 200 of them, and at least three or four times as many left hands as right. Among these are several intact prints belonging to the undamaged hands of babies only a few months old. Perhaps the most notable fact about the Gargas hands is that over 50 percent display the total loss of the upper joints of *all* four fingers. While it is certainly possible that the hands were intentionally mutilated because of some strange rite or superstition, the fact that the practice reduced the hand almost to a useless stump is rather surprising.

But were the mutilations genuine, or are they an illusion? For many years, one of the popular explanations for the painted hands was that the artists were indulging in visual trickery by folding back the joints on undamaged, healthy hands and blowing color around these distorted outlines. Certain facts now make accusations of "cheating" rather unlikely. As far as anyone can tell, the very even

3	11	1	1
9	9	1	1
3	9	1	1
1	59	1	1

A chart showing the numbers of particular types of mutilation at Gargas.

An English archaeologist, Martin Wildgoose, reproduces the sprayed appearance of the prehistoric hand impressions by blowing down a bent, perforated reed into a solution of water and red ocher.

spreads of sprayed-on color do not appear to include patches where the handprints could have been retouched or altered after the initial outline was made. More importantly, it is physically impossible to bend down *only* the top finger joint, which is missing on many of the hands. Experiments also show that it is difficult to create images as sharp as some of those at Gargas using curled-up fingers. Finally, a few of the outlines display oddly flattened or pointed ends to the shortened fingers, and these do look like mutilations rather than concealed joints.

In any case, independent evidence has recently come to light of genuinely mutilated hands in Ice Age France. In 1963, medical expert Dr. Sahly, who was conducting a detailed study of the problem, discovered a handprint with a missing little finger in the soft clay of a small circular chamber not far from the main panels of hands at Gargas. The walls and ceiling of the same chamber also were covered by rows of holes where fingers had obviously been thrust into the clay. Sahly made casts of the insides of some of these holes and revealed the shapes not of rounded fingertips but of scarred stumps, with such details visible as the lips of skin where the injury had healed together. Subsequently, several mutilated handprints were noted in clay at Lascaux, several hundred kilometers north of the Pyrenees, despite the fact that no painted impressions of such hands were seen on the walls of the cave.

The existence of people with severely damaged hands seems to be more or less proven. The key question that remains is whether these injuries were the result of intentional acts or of a crippling disease. There are no clear answers, but certainly, the restricted appearance of the mutilations, concentrated in large numbers only at Gargas, is very puzzling. If it *was* a cultural practice, it is odd that so distinctive a custom as chopping off fingers should be confined mainly to a single site and so, presumably, to a single group of hunters. The widespread stenciling of undamaged hands at so many other caves could indicate that the cultural ideas that motivated the hand paintings were not particularly exceptional but that the physical state of the Gargas subjects was. Incidentally, there are at least twenty cases of repetition at Gargas; the same hand with the same mutilation is obviously present, so that the number of people involved in making these prints may have been quite small. All these arguments tend to favor the explanation of some disease that affected only a tiny proportion of the prehistoric population, perhaps because it was short-lived or hereditary in character. Additional support for this idea may lie in the strange, twisted outlines of so many of the stumps, which perhaps indicate the remnants not of healthy but of terribly deformed fingers.

When a full list is drawn up, there are an alarming number of illnesses (most of them fortunately rare) that have the unpleasant effect of making the fingers wither or drop off. None of the specific, hereditary ailments precisely corresponds to the condition of the painted hands. Among these, Ainhum's disease, which causes the loss of the little finger through degeneration of the bone, seems to fit the case of Maltravieso, except that the disease leaves behind a stump midway up the joints and not between them, as the majority of the painted outlines seem to show.

A similar hereditary condition, Raynaud's disease, mainly attacks young women in their twenties or earlier, may be promoted by cold and damp, and is a promising explanation for the Gargas hands. This illness afflicts any of the fingers and causes them to wither through gangrene, while the thumb is not affected because of its better circulation. Once again, however, the effects do not strike neatly between each finger joint so that if either of these two infections are responsible it is necessary to imagine the prehistoric victims deliberately having the wasted portion chopped off, perhaps to discourage the spread of the gangrene. The symptoms of many of the other suggested ailments make equally unpleasant reading, and we cannot easily decide which, if any, were responsible for the mutilations. Frostbite, leprosy, and severe arthritis are among the most popular theories.

What sort of "affluent society" can it have been for these unfortunate people at Gargas, if it is true that they really were cripples, able to do little but grasp clumsily with their ravaged hands? Sahlins' ideal vision of hunting and gathering was intended mainly as a criticism of present-day values. Clearly, it provides an inadequate basis for a deep understanding of the ancient past, although it has inspired many recent romanticized accounts of life in prehistoric times. Indeed, we may feel that most moral judgments of hunters, ranging from the aggressive monster to the noble savage, have been subject to the swings of intellectual fashion, and at the moment, no universal model of the hunting past is entirely satisfactory. As the accumulation of material evidence grows in volume and detail, we have to look beyond simple preconceptions and come to terms with a past more complex and unfamiliar than was ever imagined by the philosophers of Turgot's era, two centuries ago.

8: The Human Desert

Choice and Illusion

The question of the day-to-day subsistence of hunters has provoked answers ranging from famine to affluence. Nearly as much disagreement also characterizes views of their social life. Without the help of anthropologists, we find it difficult to investigate some of these important aspects of Stone Age life—how hard it was to find a mate, for example, or how much contact occurred with people outside one's own community. Some prehistorians have simply ignored the problem and have preferred to use their imaginations. At one point, Professor François Bordes has written of the Mousterian world as "a human desert swarming with game."[1] We can readily imagine this prehistoric isolation; we can grasp what it must have been like for a few families to live together surrounded by immense "virgin" tracts of woods and grassland. But is this a correct picture?

A few decades ago, there was a measure of agreement between anthropologists about the way in which members of a hunting band were related to one another and to outsiders. With the pioneering work of experts in two continents—A.R. Radcliffe-Brown and Joseph Birdsell in Australia, Julian Steward and Elman Service in America—a coherent model of the hunting band eventually emerged. This was a local group usually composed of four or five families, for whom a nomadic way of life was essential to prevent the exhaustion of local game and vegetable resources. Many of these highly mobile bands did not move about at random, however, but within the limits of their own hunting territories. These limits were often loosely defined by stories or ancestor myths, rarely by personal ownership of the land or by privileged rights of access to resources. The band was a tightly organized, enclosed unit because it was composed of a small number of closely related people. It was also, in most cases, a male-dominated society, despite the importance of women as child-bearers or food-gatherers. It was strictly forbidden for anyone, male or female, to marry within the local group, but only the women had to leave the band in which they had grown up to join the community of their husbands. Several explanations are offered for this simplest of all kinship rules. The males of a band had to "stick together" because of their knowledge of a particular hunting territory or because of the importance of cooperative hunting tactics. Alternatively, in these small, male-dominated societies, the exchanges of women were to be regarded as "diplomatic gifts" to

Does the hunter face as empty and solitary a social life as his surroundings often suggest? Here an Australian Aborigine crosses desolate bush that has been burned off in order to drive game.

neighboring bands, strengthening social contacts and alliances over a wide area. Institutions of any real political power over and above the level of local bands were comparatively rare among hunting peoples.

This model of band society profoundly influenced a generation of anthropologists. In 1931, Radcliffe-Brown concluded that what he referred to as the "horde" was "the important local group throughout Australia."[2] The "horde" consisted of a group of males that lived and hunted together on their territorial estate, usually defined by the wanderings of a mythical ancestor creature. Spots sacred to this ancestor spirit were hallowed ground, dedicated to the observance of such ceremonies as initiation rites, and belonged to particular "hordes." Meanwhile, the women (invariably excluded from such ritual activities) were exchanged in marriage with the men of neighboring "hordes."

Despite some exceptions, this basic pattern was widely recognized among hunting societies in many other parts of the world. Indeed, in the early 1960s, Service proposed that the territorial, male-lineage band was a universal phenomenon, a basic principle of organization that belonged not only to the modern world of hunter-gatherers but also to the remote prehistoric past. Other types of social organization that had been reported by some observers in the field, involving much looser aggregations of families, were, Service declared, the result of the traditional pattern of life disintegrating under the influence of European contact.

Just as the "dismal" outlook on the economics of hunting and gathering came under fire during the late 1960s, so, too, did this tightly constricted view of the hunters' social world. Field-workers in Australia found themselves in a curious dilemma, for the "horde," as Radcliffe-Brown defined it, appeared to exist mainly in the minds of Aborigines. To this ancestral group, the aborigines owed their ritual allegiance, involving ceremonial meetings at the sacred sites, but this was not the unit of day-to-day existence. Instead, communities actually appeared to consist of families drawn from more than one of these male-lineage (or "patrilineal") groups. Moreover, in the course of their food quest, such bands would regularly range over the territories of ancestor groups to which they did not belong. The strong male influence among the Aboriginal hunting bands was not in doubt, but the idea of enclosed family groups occupying their own fixed estates was discredited.

To make things even more complicated, the existence of social ties over and above both the local band and the ceremonial ancestor group has been a subject of bitter dispute. The Australian hunters occupied a continent. They were distributed over vast regions with enormous geographical contrasts, and over this immense landscape, broad human divisions certainly existed. But attempts to define these "tribal" areas on the basis of language differences, cultural traditions, or natural frontiers have sparked fierce and prolonged controversy in Australia. Certainly, the social ties and responsibilities of many Aborigines appear much more complex than was once suspected.

Radical differences of opinion among European observers about the nature of hunter-gatherer societies were not confined to Australia. Some years before Richard Lee conducted his earliest work among the !Kung Bushmen of Dobe, another anthropologist, Lorna Marshall, had explored the !Kung communities that occupied the adjoining Nyae Nyae region. She found that the Nyae Nyae tended to congregate around permanent water holes during the dry season, which were owned by the headman of the band. This office involved the responsibility for planning the group's seasonal movements through a definite territory, and it was passed down from father to son. While the structure of the rest of the band was not a rigid one, there was nevertheless a patrilineal emphasis: the related males tended to stay together.

When Richard Lee came to work in Dobe not long afterward, however, he encountered a totally different social structure. Lee was surprised to find no trace of a headman office existing among either

Richard Lee in the middle of a cow feast among the Dobe !Kung. Several families have temporarily joined together to share the cow, and each sits in its own group, with the family head dividing the meat.

the Dobe or the Nyae Nyae !Kung hunters. Indeed, Lee discovered that the composition of each local group appeared to be utterly different to the classic male-line system described by Marshall. Instead, ownership of the dry season water hole was a communal affair, informally shared by the core of the hunting band's members. This core was composed not of related males but of sisters and brothers or male and female cousins, and its size depended on that of the water hole and the richness of the surrounding resources. The group stayed together during the leanest part of the year, and split up into smaller family units as soon as game and vegetation once more became abundant. When the season of scarcity again approached, there was a far wider choice of possible groups and water holes to join—a whole range of in-laws, cousins, and siblings—than could ever be possible under a strictly male-lineage organization. In other words, the fact that !Kung society was not, in kinship terms, male-dominated meant that there was a great deal of flexibility and choice in people's lives. Adhering to the rule of a restricted, territorial male line might be dangerous or even suicidal in the desert, where during the drought it is so hard to predict where food is to be found from one week to the next.

There has been an even more remarkable development in the study of the Kalahari hunters and gatherers. The findings of two of the latest field-workers, John Yellen and Henry Harpending, report not merely flexible and fluid interband relationships but a lack of any recognizable system at all. The movements of the desert hunters are so unstable from one season to the next that no rules apply to the way families congregate and disperse. According to their report:

> the !Kung band or local group may be regarded as a temporary and unstable aggregation of individuals and nuclear families

which, over the long run, may be viewed as moving randomly over the landscape. Changes of personnel of over 50% have been noted in the course of a few years. . . .[3]

As confirmation of this new picture of the extreme social mobility of Bushman life, genetic tests have revealed a remarkably low factor of inbreeding, much lower than is present among most settled agricultural populations of today.

To combat the unpredictable pressures of the desert, the !Kung move far and frequently among a wide range of relations. Such movements do not only have the effect of evening out scarce resources over a large area but also help to lessen social tensions. Numerous relatives camping around a water hole during the drought may mean security, but people have to work harder to feed more people, and social temperatures run higher. As Lee explains:

> the Bushmen sought both kinds of social existence, the intensity and excitement of a larger grouping and its attendant risks, and the domestic tranquility and leisure time of smaller groupings.[4]

And if the social intensity of a large dry season group boils over into a serious dispute, it is usually possible for a !Kung family to pack up and move to another water hole, where other relatives are in residence.

In Australian Aboriginal society, personal freedom is narrowed by elaborate marriage systems and rules, moral codes more rigid and restrictive than exist in any Western society. Such involved social arrangements create problems for the individual, often making it impossible to find a suitable mate. Yet because of their very complexity, these systems do connect together families that are remote—remote in terms of both blood relationships and, often, geographical separation.

Moreover, in both the Australian and Kalahari deserts, common kin names are applied to categories of people regardless of how close or distant their actual relationship may be. This places individuals who may happen to be only indirectly related to the members of a hunting band (and who have perhaps traveled across the desert from a long way off) on an even footing with the immediate local kinfolk of the group. Virtual strangers, then, are rarely outcasts in hunting societies. Social mechanisms exist to link together populations that are widely dispersed and constantly on the move.

In many communities the options are opened still further, beyond the facts of birth and blood relationship. Ties are formed that have no basis in family history at all, but are sanctioned by ancient customs or myths. No environment could possibly resemble the Kalahari less than the Canadian wastelands north of Hudson Bay; yet in the bleak Arctic wilderness, social tactics for dealing with the problem of unpredictable food sources are surprisingly close to those of the desert.

For example, during the summer and late autumn, the population of the Netsilik Eskimo fragmented into small family groups, mainly in order to exploit the salmon runs at specially constructed stone weirs. When migrating caribou appeared at the fording points along rivers in the early fall, larger bands of hunters formed to cooperate in driving and spearing the animals. However, the real focus of Netsilik life was the arduous winter, when the only substantial resource available was seal meat. The task of spearing seals through their breathing holes in the ice was a risky affair. One or two male hunters, providing for a single family by a constant watch at a breathing hole, would need incredible luck to secure a steady catch of seals. But a large number of hunters deployed at many points across the ice, all sharing the resulting catches, reduced the risk substantially. Consequently, during the winter the Netsilik congregated into settled igloo villages containing sixty people or more, which encouraged intense social and ceremonial activity of all kinds.

In these relatively secure communities there were precise rules which applied to the sharing of seal meat. These rules applied not only to one's immediate relations but also to sharing partners who had no blood relationship at all. During childhood a mother would select roughly a dozen such partners for her son, each of them permanently appointed to share in a particular part of the seal's anatomy. So, when the hunter brought his catch back to the settlement, each of these partners was entitled to a specific portion of the corpse. A recent researcher among the Netsilik, Asen Balicki, explains that special names were also involved in this division of the carcass:

> Meat-sharing partners called each other not by their personal names but by the corresponding seal cuts they exchanged. If "A" and "B" exchanged seal buttocks, they called themselves reciprocally "UKATIGA: my buttocks." They considered each other close friends, almost like relatives. In case non-relatives without sharing partners were in camp for the winter season, they were integrated into the sharing system by selecting their own sharing partners.[5]

A rigid enforcement of family-centered relationships and duties would be as disastrous in the cold desert as in the hot. Mobility and flexibility were essential to survival, and "artificial" devices like that of the sharing partner provided additional, alternative ties to those of the family. Such factors may explain the apparently casual attitudes to the suffering of close relatives noted in the last chapter. The uncertain forces of nature tended to shatter and fragment the human community, dispersing it into perpetual restless movement. Yet opposed to these divisive forces were all kinds of social inventions which had the effect of binding people together in a wide variety of relationships. Choice always existed in the hunter's social world, even if nature offered few options.

Because some anthropologists believe that the movements and group structure of a people like the !Kung are random and chaotic, it would be wrong to assume that disorder and novelty constantly upset the ideas and traditions of the hunter-gatherer. Quite the reverse appears to be true. One of the most interesting aspects of recent archaeological work in desert regions is that it has shown how extraordinarily conservative cultural traditions and patterns of behavior are in such environments.

For example, the remains of late Stone Age campsites in the northern Kalahari appear to fall into a pattern of location very similar to that of the present-day !Kung. Preliminary investigations of these sites seem to indicate a slightly more arid climate than today, but much the same types of animals were hunted. The Stone Age tool kits (of the so-called "Wilton" culture) show a remarkable uniformity from site to site across long distances, suggesting that the prehistoric hunting bands of the Kalahari maintained extremely fluid social contacts just like the Bushmen of recent times.

Similar investigations in the arid southwest regions of North America and in the Australian deserts have clearly demonstrated that patterns of existence persisted almost unchanged over the course of thousands of years. Richard Gould's recent excavation of a rockshelter site in the Australian Western Desert revealed successive layers of material scarcely different from that used by the Aboriginal group which inhabits the area today, yet the timespan covered by the dig was no less than 10,000 years.

There is little place for innovation in the desert. Unchanging traditions, whether in the form of material equipment or age-old ancestor myths and rituals, are another way in which hunters express their solidarity in the face of constant insecurity and movement. Indeed, the "timeless" notions of the hunters may occasionally conflict with the actual flexibility of their behavior. This makes the task of the anthropologist particularly difficult, especially if he is only able to stay among the band for a brief period. At least part of the confusion over the social structure of hunting bands may have arisen because of these contrasts between belief and action: traditional values that pass unaltered from generation to generation, and social arrangements that change from season to season.

The environment of Ice Age Europe was not a desert. The extreme pressures exerted on communities dependent for survival on moist roots and bulbs or on breathing holes through the ice are likely to have been rare experiences for Mousterian or Magdalenian hunters. Yet the difference is one of degree, not kind. A Magdalenian band surrounded by seasonal abundances of reindeer and salmon faced problems similar to those of the Netsilik, even if they could be solved more easily. Without social arrangements that tended to iron out the irregularities of scarcity and abundance, individual Magdalenians could have faced hard times as surely as a Netsilik without a sharing partner. Only a careful examination of the archaeological evidence can provide a detailed picture of the kind of social life which the prehistoric hunters led. Are there signs of the constant dislocations, the far-flung family contacts and the cultural conservatism, which are seen in their most extreme form among the desert hunters?

The Units of Society

The existence of the family in Paleolithic times is scarcely open to doubt. The type of man-made structure most commonly excavated is a small hut about four to six meters across, with a single hearth at the center. The frequency of structures of this size—repeated in every variety of shape and building material imaginable not only in France but in central and eastern Europe as well—strongly suggests that the family was the fundamental unit of society.

Beyond this simple conclusion, however, the remains of hearths and buildings do not take us very far. It might be naïve, for example, to assume that a big hearth built under a cave porch helped to feed and keep warm more people than a small one. More useful information comes from the hunting settlements of Moravia and the Ukraine, where the remains of artificial structures are preserved, as at the famous sites of Dolní Věstonice and Kostienki. While several of these settlements comprise four or five separate huts, it is usually difficult to estimate the number of inhabitants at any one time, for it has often proved impossible to know if all the structures at a single site were contemporary.

Occasionally, larger buildings appear to have taken the place of the individual family units. The discovery of the first of these "long houses" at Kostienki on the Don in 1879, later excavated over the course of thirty years, encouraged much speculation among Soviet scholars. The Kostienki I house was eventually reconstructed as a large rectangular communal building, heated by a long line of central fires. The framework was probably built of wood, resting on low banks of clay, and covered an area about eight meters wide and almost thirty meters long. The size of this enclosed space encouraged some Russian researchers to speak of a "clan settlement," corresponding to an imaginary type of communal society that supposedly existed before the growth of the conjugal family.

However, later discoveries of other long houses at Kostienki and at Pushkari showed that some of these large structures could just as plausibly be interpreted as composite dwellings where several basic family-sized units had been joined together in a line. A mundane explanation for these communal houses may be that it was easier or more economical to build and heat one structure than several detached huts. In more recent times, groups of Siberian families have often collaborated in the erection of long houses for settled wintertime occupation. The quantity of flint and bone debris at several prehistoric sites in the Ukraine suggests that many years of successive occupation may be represented by the remains. We can imagine the same families returning to live together winter after winter in these substantial houses; like the wintertime sealing villages of the Netsilik, the wooden long houses may have provided the focal point of prehistoric social activity and ceremony during the year.

Even these large buildings may be unreliable guides to the number of people in a hunting band. The investigations of complicated interlinked houses at Barca, in Czechoslovakia, showed that different activities were carried out in separate parts of a single structure. In

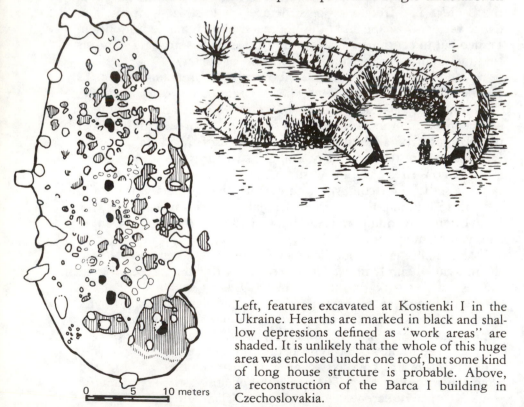

0   5   10 meters

Left, features excavated at Kostienki I in the Ukraine. Hearths are marked in black and shallow depressions defined as "work areas" are shaded. It is unlikely that the whole of this huge area was enclosed under one roof, but some kind of long house structure is probable. Above, a reconstruction of the Barca I building in Czechoslovakia.

this case, a crude estimate of the number of inhabitants based on the available floor space might be misleading. Certainly the effort involved simply in gathering construction material (especially heavy mammoth bones) at several of the settlements in the river valleys of the Ukraine does suggest that many hands were available at certain times of the year.

Some of the most famous and heavily occupied rock shelters in the west—La Vache and Duruthy in the Pyrenees, for example, or Laugerie Basse and La Madeleine in the Périgord—probably represent the natural equivalent to the man-made structures of the east. A few of these deposits are so extensive (one of the shelters at Laugerie extends for 180 meters) that one can envisage groups of several dozen families occupying the porches at any one time. Away from the shelter of the caves, large open-air settlements were created that may have resembled those of Moravia and the Ukraine.

For several years a joint American and French team has been striving to expose and record a series of enormous cobbled pavements, successively laid down beside the Isle river at a place called Solvieux, some fifty kilometers from the Dordogne. Because of unfavorable soil conditions, bone remains and the traces of structures are not nearly so well preserved as in the river terraces of eastern Europe, and as yet it has proved difficult for the researchers to recognize the outlines of individual buildings. However, Solvieux must have accommodated a great number of people. One floor that has been exposed stretches for a full fifty meters; another, much later, pavement covers at least one hectare, and is littered with thousands of flint tools. Cobblestones cracked and reddened by fire are scattered everywhere. These probably represent the aftermath of cooking activities or of efforts to soften frozen ground. As the initial report of the archaeologists concludes, the Solvieux pavements "clearly represent many people doing many things over substantial periods of time."

In the same valley as this great settlement, intensive survey work has revealed the presence of over twenty other sites, some of them apparently single and isolated hut floors only a few meters across. The contrast in size among the Isle valley sites and elsewhere suggests the kind of fluctuations and seasonal movements characteristic of the !Kung and many other recent hunting peoples. During temporary periods of either abundance or insecurity, large groups of families banded together in cave porches and substantial structures. At other times of the year they dispersed into smaller units—each family perhaps going its own separate way and camping in tents or huts that could be speedily erected and dismantled, or left intact for the next year's visit.

The natural seasonal movements of animals probably encouraged many hunting groups to congregate and then disperse in a regular annual pattern. As we have already seen, it is in exactly this way that the late Ice Age sites of Germany seem to reflect the constant movement to and fro of wild reindeer herds. If Sturdy's ideas are correct, the policy of exploiting the reindeer by discreet, indirect contact imposed a migration cycle on the human group that corresponded to the instinctive wanderings of the animals. There were fixed summer and winter bases at either end of the migration trail, but during the great spring and autumn treks the human herd followers must have taken to the wilds and camped in tents or huts.

The Great Migrations

Among the most intriguing conclusions of all to emerge from recent Ice Age research is that this regular pattern of seasonal movement seems to have happened not only in Germany but also among the communities of Magdalenian hunters and cave artists in the Dordogne and Pyrenees. Some of the evidence for constant mobility and widespread contacts has been well known for years (the uniformity of style in many Magdalenian carvings and paintings has always suggested this). Only recently, however, have serious attempts been made to pull all the scraps of evidence together in a coherent picture of the way of life led by the reindeer hunters. Two English scholars, Ann Sieveking and Paul Bahn, have separately drawn attention to many neglected clues which all point to the fact that prehistoric people in France undertook regular long-distance travels.

The intimate connection between the Périgord and the Pyrenees is most obvious in the small carved pieces of bone and antler, most of which were dug up from the cave entrances during the last century. Such objects, which range from the mysterious batons to practical hunting weapons like spear-throwers, begin to be heavily embellished with engravings during the middle of the Magdalenian period. These engravings display unmistakable similarities of style which appear to span wide geographical distances. There are animal carvings from Isturitz in the western Pyrenees, for example, which show identical conventions of perspective and shading to others from La Madeleine in the Dordogne.

Consider the supposed "bridle" pattern which appears most frequently on the muzzles of little cut-out bone horse heads at Pyrenean sites like Mas d'Azil and Arudy. Whether or not we think of these as representing bridles, it is remarkable that this identical pattern should appear on an engraving of horse heads from Laugerie Basse in the Dordogne and on the famous bison spear-thrower of La Madeleine. More general similarities of design closely link bone objects such as disks, spear-throwers, and spatulas engraved with fish, discovered many hundreds of kilometers apart. The most remarkable case

(above) The celebrated carving of a bison from La Madeleine in the Dordogne that probably served as the weighted end of a spear-thrower, and (*below,* at a different scale) another carved bison from Isturitz in the western Pyrenees. Together with the horse heads from the central Pyrenees (see page 110), these widely separated carvings all display similar features of style, such as the peculiar "bridle" pattern on the muzzle. The presence of this pattern on the two bison is, incidentally, a good argument against the belief that it really does represent a bridle.

is that of bone objects from the cave of Kesslerloch in the northern Swiss Alps which are decorated in a style indistinguishable from those of the Dordogne, about 450 kilometers away.

The obvious explanation for these similarities is that a very small number of highly skilled artists was responsible for the production of the carvings. This idea is strongly supported by the fact that in both the Dordogne and the Pyrenees carved objects have been found in large numbers at only a few great sites. If every hunting group supported its own specialist sculptor, then we would expect the objects to be more evenly spread throughout both regions. As it is, one expert has calculated that 66 percent of all the mobile art pieces of the Périgord come from only four sites. Once finished, however, the products of these craftsmen were in widespread demand, for they appear to have been carried far across Ice Age France.

Decorated objects were not the only special materials to travel over considerable distances. Besides some startling resemblances in certain tool kits discovered in both the Pyrenean and Périgord regions, caches of fine unworked flint blades have been discovered in the entrances to several caves in the foothills of the Pyrenees. This is particularly interesting because Pyrenean flint is scarce and of such inferior quality compared to that of the Dordogne that it is unsuitable for the manufacture of the blade type of Paleolithic tools. Other objects made of substances not locally available have also been noted at a few Pyrenean sites. Someone, then, was moving objects and materials to and from the caves of the foothills, and while specialist craftsmen or traders may well have been involved, why should we not imagine whole populations of hunters on the move as well?

Only the laborious examination of reindeer antlers and salmon bones can provide reliable indication of the season at which a cave site was occupied, and few thorough studies of this kind have yet been published. The facts which are available show that certain important caves in the Pyrenees, including Isturitz, Duruthy, Gourdan and Lespugue, were wintertime bases during the Magdalenian period.

Most of these sites are dominated by the remains of reindeer, sometimes in extraordinary numbers. At Gourdan, in the central Pyrenees, Piette estimated that he had dug up the bones of at least three thousand reindeer, and the deep layers of soil and stone suggest a long period of successive winter settlements at the site. The location of most of the caves or shelters enables anyone standing near the archaeological deposit to obtain a commanding view over the narrow steep-sided valleys. This position may have been useful for keeping a watch on the grazing of herds in the flat bottoms of the valleys which were worn smooth by the passage of ancient glaciers. Indeed, the shape of these "bottleneck" valleys suggests that they would have been ideal for the indirect methods of reindeer herding described in Sturdy's study of the German sites. As winter approached, we can imagine the Magdalenians arriving in the wake of their huge reindeer herds, while a group of hunters busied themselves with dividing off a portion of the herd, which was then driven into the confined spaces of the foothills within sight of the old home base. Then perhaps a great autumn slaughter and feast would take place or else, more prudently, the hunters would simply make efforts each day to ensure that the portion of the herd on which they were relying for the whole winter did not stray too far from their view at the cave porch. The question is, what happened as spring and summer arrived?

The simple answer, which would explain why nearly identical art objects are found in the Dordogne and the Pyrenees, is that they traveled north with the herds and settled in the great shelters of the Dordogne and Vézère such as La Madeleine and Laugerie Basse. In other words, the reason for the similarities between sites separated by hundreds of kilometers could be the same group of people settling in both regions at different times of the year.

This, however, is a very unlikely solution. A new survey of the deposits which cover the valley bottoms of the Périgord has shown that throughout Magdalenian times there were intense natural forces at work in the region. Frost shattered the rock faces into tumbling scree, while strong water erosion washed soil down from the surrounding plateau country. In early spring and summer these processes would have transformed the valley bottom into a waterlogged, muddy morass alive with insects. This was the last sort of place to which either reindeer or man would have migrated in the summer. Too few studies have yet been undertaken to permit a firm, independent conclusion about the season during which most of the caves in the Dordogne were occupied. At the moment it does seem likely that they were winter bases like the Pyrenean caves to the south. Therefore we have the problem of understanding how two distinct groups of people, who settled for the winter in different regions, came to share the same art traditions and to exchange certain raw materials. The obvious answer is that they met during the summer, but where?

There are various possibilities. The reindeer herds could have wandered up to cool mountain pastures in the east, and this might explain why carvings in unmistakably Magdalenian style were found high in the Swiss foothills at Kesslerloch. It is even possible that some of the reindeer which wintered in the Pyrenees simply moved a short distance higher up in the same mountains, although the terrain to the south of many sites rapidly becomes extremely steep and rugged. It is more likely that as the calving season drew near, the does behaved as Arctic caribou of the Northern Territory do today, and massed in considerable numbers on flat ground close to the fresh breezes of the ocean. If this assumption is correct, it would mean that in spring some of the hunting bands from both the Pyrenees and the Périgord may have set out toward the flat plains of the west bordering the Atlantic while others headed south to the Mediterranean. Seaweed on the beach would have provided ideal summer browsing for the reindeer. Can it be proved that the hunters were there, too?

It is indeed unfortunate that the shorelines of prehistoric times now lie under about one hundred meters of sea water. The only evidence for possible summer expeditions to the coast must come indirectly from the objects which the hunters brought back to their winter shelters far inland. A favorite item for collection seems to have been small spiral seashells, mainly of globular or elongated shapes, which are not the easiest to find, but are among the most decorative. These shells are found in their hundreds in cave deposits and burials, and are usually perforated for suspension as parts of necklaces, bracelets or dress ornaments.

The modern habitats of these mollusks may have shifted considerably because of warmer sea temperature since the Ice Age, so we cannot assume that particular species were definitely collected either beside the Atlantic or the Mediterranean. It is reasonable, however, to think that shellfish which can only tolerate seas of Mediterranean warmth today certainly could not have been present in the chilly Atlantic waters of prehistoric times. On this basis, it is apparent that mollusks from both coastlines are represented in the finds from Magdalenian sites. As we would expect, the sites in the Dordogne that were relatively close to the Atlantic show a high proportion of hardy species in their deposits, while the communities of the central Pyrenees had greater access to those mollusks limited to the Mediterranean. Taken at face value, this would suggest that visits to both coastlines were undertaken from all the regions of southwest France, although the importance of the Atlantic shells suggests that this area was usually the more common destination, even for the hunters of the Pyrenees.

A crucial objection to the theory of summer migration to the sea is the possibility that the shells were carried inland through some indirect exchange network, perhaps as "shell money" bartered for reindeer meat or hides by far-off communities whose coastal bases are now lost to us. However, there are other marine objects in Magdalenian deposits that seem rather unlikely items of trade, including the jaw of a harp seal found at Raymonden in the Dordogne, about two hundred kilometers from the present shoreline. Seals have been observed to make occasional forays inland up major rivers in pursuit of salmon, so it is conceivable that seals could have died locally in the Dordogne. However, such an explanation can hardly apply to a poorly preserved fragment of sperm whale tooth fashioned into a relief of two ibexes that was discovered in the great tunnel cave of Le Mas d'Azil in the central Pyrenees.

There are a few artistic representations of marine animals from other sites in the same region, notably seal engravings from Gourdan and La Vache and a carved bone silhouette of a flatfish from Lespugue. Apart from the unusual subjects depicted, there is nothing special about these decorated objects which might suggest that they were traded from afar. At the same time, the fact that depictions of marine fish are scarce compared to the many engravings of freshwater and continental species (particularly those of the salmon family)

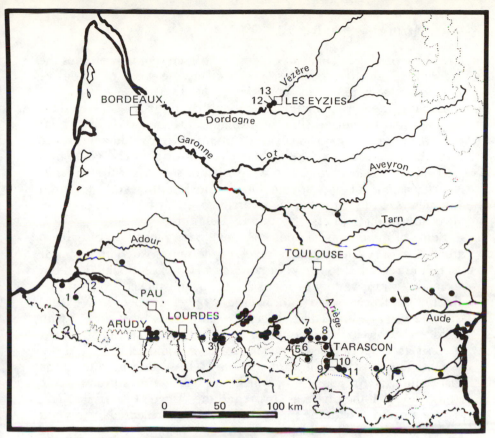

Magdalenian occupation sites in the Pyrenees. In the Dordogne region only
the two best-known sites have been marked. The 1,000-meter contour is
shown. 1. Isturitz, 2. Duruthy and nearby sites, 3. Labastide, 4. Enlène, 5.
Les Trois Frères, 6. Le Tuc d'Audobert, 7. Le Mas d'Azil, 8. Le Portel, 9. La
Vache, 10. Les Eglises, 11. Fontanet, 12. Laugerie sites, 13. La Madeleine.

does suggest that fishing activities were mainly confined to riverbanks and estuaries. The most economical explanation for all these
facts is that Magdalenians from the Périgord and the Pyrenees did,
in fact, visit the coast, not primarily to supplement their diet by sea
fishing or beachcombing, but in order to keep track of the movements of their reindeer herds. The seasonal travels of the hunters
may account for many of the similarities among engraved art objects
found throughout Paleolithic France.

The pattern of movement and exchange was not a simple one: the
people of the Périgord and the Pyrenees did not share everything.
A few important categories of decorative bonework are found only
in the southern sites, including a whimsical spear-thrower design of
a defecating fawn. This was evidently in vogue, since some five or
six copies of it are known from sites widely spread across the foothills, and yet so far no examples of it are known from the Dordogne
region. Other motifs appear to be concentrated either in the western
or the eastern group of sites in the Pyrenees. All this suggests that

the range of Magdalenian art objects from each region preserves the traces of distinctive local tastes as well as the signs of extensive contacts elsewhere.

Equally, there is fascinating evidence that individuals traveled far in ways that cannot possibly be explained as the result of regular seasonal migrations. Fossil curios of an attractive or odd appearance were widely distributed from their place of origin. The geological beds of the lower Loire region were a major source for these fossils, which passed not only to the Dordogne region several hundred kilometers to the south, but were carried even farther in the opposite direction to the hunters of southern Belgium.

Some of these unusual finds may have been picked up casually in the course of regular seasonal travel, yet certain isolated finds appear to be the record of journeys or contacts of an exceptional kind. For example, a rare variety of fossil mollusk shell picked up from the floor of the Lascaux cave is known today only from beds in Wexford and the Isle of Man. This particular example has been carved with a slot, probably so that it could be threaded onto a string. Similarly, a fossil at Laugerie Basse appears to have come from the Isle of Wight. The most striking case concerns a fossil dug up during the last century at Arcy-sur-Cure, which was considered so remarkable at the time that the cave was christened the "Cave of the Trilobite." The nearest source for the species of trilobite represented by this find is about two thousand kilometers to the east in central Germany. The problem with all these fossil discoveries is that sources nearer to hand may have existed in Paleolithic times which are unknown today (perhaps because they are now under the sea).

However, the existence of far-flung contacts is hardly to be doubted. Common items of Magdalenian bonework found their way into the possession of many late Ice Age hunting groups throughout Europe. We cannot know exactly how harpoons of a typically Magdalenian form came to be at Creswell Crags, Derbyshire, toward the end of this period, nor why finds of the perforated batons reach from Tito Bustillo in the northwest of Spain as far as Mezin in the Russian Ukraine. Perhaps the most interesting question of all is how a remarkably uniform art style came to be practiced not only in the decorated caves of southern France, but in northern Spain as well.

The hunter-artists of the Périgord and the Pyrenees probably had a natural opportunity each year to meet and renew contacts in the course of their seasonal travel, but how their traditions spread to Cantabria is not nearly so obvious. Economic factors can scarcely have bound these regions together. No large herd of reindeer would ever have migrated in spring from the north side of the Pyrenees over the cool mountains and down into the humid coastal belt of Cantabria. Besides, in northern Spain itself we know that the Magdalenians were practicing a quite different economy based mainly on

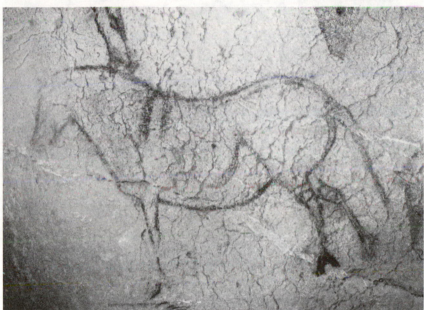

The famous black-painted "circus horse" of Le Portel in the central Pyrenees and *(below)* one of the many fine horses at Ekain in the Spanish Basque
region. These Magdalenian paintings share many general features of style.

red deer and ibex, and probably involving seasonal displacements of
fairly short distances between the Cantabrian mountains and the
coast. To account for the striking resemblances between cave art in
France and Spain, there must have been close contact arising mainly
from social rather than economic necessity.

The evidence for artistic contact will not fail to impress the tourist
who, for instance, visits Niaux in the Pyrenees and then Santimamiñe

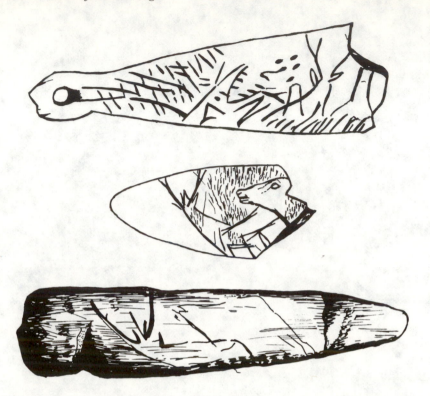

The "wolf and deer" theme as it found expression on bone objects from *(top to bottom)* El Pendo, northern Spain, Lortet and Le Mas d'Azil, central Pyrenees.

in the Spanish Basque country. Much of this Magdalenian engraving and painting looks and "feels" the same, as if specific conventions of style and theme were closely observed. There are uncanny correspondences in the details of shading and coat markings (especially in pictures of horses) between the two regions. Indeed, when confronted by the prancing "circus horse" of Le Portel in the Pyrenees, one naturally wonders if the same master hand was not also responsible for paintings of an almost identical quality at Ekain, over three hundred kilometers away in the Basque country. Fortunately we do not have to rely on personal reactions like these to assess the close relationship between the two areas. A curious and rare design which appears to show a wolf's and a deer's head side by side has been found engraved on three bone objects, two of them from the Pyrenees and the third from the cave of El Pendo near Santander in northern Spain. Other, more common items of bonework, such as knives with fish designs, also suggest a very close link between the late Ice Age hunters of France and Spain. This relationship may have been founded on a network of marriage ties between the two regions or reinforced by other institutions which could have promoted the exchange of objects or the movement of people.

Many have assumed, quite mistakenly, that the coming of agriculture and a settled existence were essential to encourage the development of specialist miners, peddlers and artisans. In fact, objects moved in the Australian desert partly because there were Aboriginal traders who traveled great distances to seek and supply desirable items for exchange. These items included heavy grindstones, shells, ocher, the narcotic leaves of the *pituri* plant, and hunting weapons, which were frequently carried far across inhospitable terrain. Records left by observers in the last century leave no doubt that expeditions of eight or nine hundred kilometers in search of *pituri* and ocher were by no means exceptional in desert regions. The men of one tribe, the Dieri, are recorded as occupying two months with the extraction of ocher, transported in string bags which weighed up to seventy pounds each. Other accounts speak of Aboriginal women carrying massive grindstones on their backs for considerable distances. These records indicate that even in a desert highly mobile individuals or small groups could support themselves for long periods while performing special tasks. If we believe that prehistoric societies were as complex as those of the Aborigines, then the possible existence of seashell merchants or individual cave artists who decorated one cave in the Pyrenees and another in the Spanish Basque country is entirely plausible.

In Australia craft specialization also extended to finished products such as shields, spears and boomerangs. The ethnographers Baldwin Spencer and F. J. Gillen, whose classic work on the tribes of central Australia was published in 1899, record that the trade in these objects was not related to obvious supplies of raw material, but to the traditions of excellence in design or finish for which particular tribes became famous. Such reputations covered great distances, as Spencer and Gillen commented:

> Every Arunta man is sure to have one of these shields, and yet the majority of them have not been made in the tribe, nor, indeed, within a hundred miles of the district occupied by it, but by a tribe speaking a quite different language.[6]

In 1898 another ethnographer recorded seeing a special type of hooked boomerang over 1,200 kilometers from its place of manufacture. The transmission of such objects over long distances was not only due to the movements of specialists but also to trading activities between individuals, both at the level of personal gift-giving, and in the extremely formal setting of religious gatherings. This is how one modern anthropologist of the Western Desert, Ronald Berndt, has recently described such gatherings among the Walmadjeri and Gugadja peoples:

When large ceremonies and rituals are held, some of the participants come from places a great distance apart; they provide,
therefore, an ideal opportunity for bartering. Trade takes place
within the context of ritual and often is not seen as being
something separate. Items that change hands at such times include red ochre, spears and native tobacco. . . . But the most
important goods are pearl shells and sacred boards.[7]

So important, in fact, were certain distinctive shells that they found
their way from one side of Australia to another. A recently published
map shows that Aboriginal groups passed at least one type of shell,
used in circumcision ceremonies, from its origin on the northern
coastlines of the Barrier Reef and Carpentaria right across the desert
heartland of Australia and into the far west and south of the continent. A distinction between practical objects of economic value and

The intense ritual and social life of the Australian Aborigines is centered
around night-time gatherings, or corroborees, such as this one photographed in 1974.

Aborigine hunters on the lookout for game in Arnhem Land, Northern Territory, in 1977.

objects in the realm of art and religion did not exist for the Aborigines, and so the traffic in sacred shells or carvings was no more or less important than the exchange of spears and boomerangs. Whatever ritual beliefs or acts were involved in the circulation of Magdalenian carvings or the decoration of cave walls are likely to have involved contacts and movements as essential to Paleolithic people as the practical demands of seasonal migration and herd following.

Perhaps it is unwise to push the comparison of Magdalenians and particular desert hunters of recent times too far. It is the general aspect of their social lives which may strike us as similar. There is clear evidence that just like the !Kung or the Netsilik, the cave artist communities were extremely mobile, not only physically but in social terms as well. No single theory—whether it be in terms of regular seasonal movement, specialist traders, marriage ties or religious ceremonies—can satisfactorily explain all the evidence; no one activity could possibly result in the complex patterns in which the cave paintings, engraved objects, shells and flint tools are distributed. Extremely fluid social relations must have existed, otherwise we cannot explain why art styles were closely followed over wide areas, nor why the practice of decorating caves with animal pictures endured for 25,000 years.

It is true that Paleolithic flint equipment did change over this period of time, unlike the almost incredible continuity which archaeologists have revealed in the desert tool kits. It is also likely that reindeer and fishing resources ensured a more regular and secure existence for the French Magdalenians than the hunters of the Arctic Circle or the Kalahari have known in recent times. If the grassy plains and parkland of the Périgord were in some senses a desert, it was not a forbidding or harsh one. Despite these contrasts, social flexibility and timeless traditions have shaped the lives of hunters past and present.

9: *Footprints in the Sand*

Cave Art—Public and Private

Leaving the spectacular cliffs of the Dordogne valley behind and heading south along the minor roads, a tourist soon finds himself on an arid plateau crisscrossed by narrow lanes and drystone walls. This is the upland "causse" country of Quercy, where the thin soil supports a light cover of birch trees, juniper bushes and coarse grasses, attracting thousands of bright butterflies in the summer. It is possible to walk only a hundred yards among the birches and become utterly lost in the featureless scrub, with the distant clatter of sheep bells in every direction confusing the senses further. It is a kind of desert, in fact, which one can scarcely imagine ever to have been prosperous or continuously inhabited. Yet the signs are there: in the clearing around a field boundary here and there loom the massive stones of a megalithic tomb, or "dolmen," in all probability erected by Neolithic shepherds in their summer pasturelands four thousand years ago, and surprisingly there are also the painted caves left behind by the Ice Age hunters.

It was in the middle of one such stand of birches, a place impossible to find without a guide, that the cave of Les Fieux was discovered by a group of potholers in 1964. No impressive cliff or porch confronted the explorers, for the entrance was a mere "foxhole" disappearing into the earth beside a shallow depression in the flat stony plateau. In prehistoric times this narrow crack opened onto a broad chamber with a thin roof shattered bit by bit by winter frosts and

Two intact left hand impressions may be seen faintly in this photograph taken in the cave of Les Fieux in the Quercy region of southwest France.

eventually filled up with sand and the hunters' domestic rubbish. On the surface the only trace to be seen of this was the shallow depression. In the other direction the crack led down into the darkness of the Les Fieux cave.

Monsieur Caminade, the landowner, a vigorous man in his eighties, has always led visitors himself into the depths of his cave despite the fact that his powerful shoulders can barely pass through the narrow "foxhole" of the entrance. Crawling along on elbows and knees, one descends from the hot plateau into the coolness of the cave, and at length scrambles into a rather dingy chamber about twenty-five meters long and thirteen meters wide. It is no Lascaux or Altamira: some of the dozen or so painted hand impressions are very hard to see, the triple red dots made by dabbing at the cave walls with three fingers are uninformative, while the only other traces are some extremely worn carvings, apparently representing a pair of ibexes, covering one side of a block lying in the middle of the chamber. It is indeed hard to imagine a great ceremony or gathering taking place here in this claustrophobic chamber where one slides and stumbles among muddy little crevasses and can easily be knocked painfully across the head by bristling stalactites.

Les Fieux would not command much attention if it were just an isolated case of a small and poorly decorated sanctuary. In fact, as a result of careful prospecting during the past decade, nearly a dozen caves, their locations dotted over the dry hills and steep river valleys of Quercy, have been found to contain unimposing paintings or engravings. The caves are never more than a few hours' walk from the great rivers, the Lot, the Célé and the Dordogne, but to some extent these recent discoveries fill in blank spaces on the map between caves like Font-de-Gaume near the Vézère and Pech-Merle on the Lot which have long been celebrated tourist attractions.

However, no enterprising proprietor could ever turn these new caverns into profitable concerns, for who would pay to see Le Bourgnetou with its almost invisible reindeer carving and its solitary red handprint, or Les Escabasses with its ibex and two sketchy horses hidden among ugly walls of spongy calcite? It has been relatively easy for prehistorians to erect grand theories about the significance of caves like Lascaux where hundreds or even thousands of artworks survive. But it is difficult to apply the same ideas to the impoverished scale of decorations in caves like Cuzoul de Mélanie (one engraved bison) or Le Papetier (one engraved bovid and three abstract signs). Yet these recent finds demand an explanation no less than did the dramatic discoveries of the past. Perhaps they suggest the activities of solitary artists rather than the settings of great ceremonies.

Within cave systems of considerable size where we could imagine prehistoric gatherings of several dozen or even hundreds of people,

there are tiny painted cavities—"private chapels" one might suppose
—capable of admitting only one or two people at a time. One of the
most striking examples is in the cave of Tibiran in the Pyrenees,
where paintings and engravings were first noted by the celebrated
explorer Norbert Casteret in 1951. From the entrance one slithers
alarmingly down a steep muddy slope to the central chamber below.
This room is a high-vaulted amphitheater about twelve meters wide,
complete with an "upper circle," a high ledge where prehistoric
artists left mutilated hand impressions on the wall (Gargas is only a
short walk away on the other side of the same hill). The theatrical
atmosphere of this chamber is strong, readily suggesting a place of
dance or sacrifice to the imaginative. Yet the most interesting works
in Tibiran occur not here but in a little cavity joined to the main
chamber by an extremely low opening in the rock wall.

One slides painfully on hands and knees into the tiny chamber,
where there is just room for two people to stand upright. One
careless step, a hand flung out to catch one's balance, and the already
fragmentary animal paintings could be seriously damaged. The best
of these is a vigorous black painted horse, its outline engraved on
two separate occasions, while on the adjoining wall a drooping stalactite shaped exactly like a bear's head is joined to a painted body, as
if the artist was inspired by the natural relief of the stone. There are
also traces of an ibex and possibly another horse. The idea of a large
ceremonial gathering in the central chamber may of course be just
a fantasy, but it is absolutely certain that not more than two people
could ever have squeezed into the cavity with the painted animals at
one time.

This apparently deliberate selection of an awkward secret place
has emerged also from a recent study by two English cave-art enthusiasts, David Collison and Alex Hooper. Together they spent six
weeks in 1972 minutely examining the confined walls and ceiling of
one small gallery inside a cavern opening onto the spectacular rocky
slopes of the Ariège valley, not far from Andorra in the Pyrenees.
This cave of Les Eglises had been the successive shelter not only of
Paleolithic hunters but also of Bronze Age peoples whose fragments
of pottery have been excavated from the entrance, and, in the thir-
teenth century, a sect called the Cathars took refuge from religious
persecution in the upper galleries of the cave.

Nearly 100 meters from the entrance and not entirely out of reach
of daylight is a small side turning that leads up to the dead end of
the painted gallery, where the semicircular roof and walls form a
smooth and continuous surface joining the gravel, clay and sand
floor. In its final sections the tunnel becomes lower and narrower so
that one cannot stand upright, and it is here in this cramped section
that there are clustered eighty rather unimpressive red-painted and

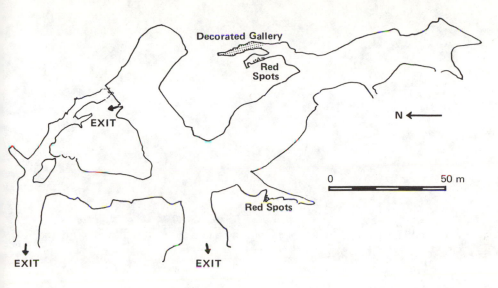

Plan of the decorated cave at Les Eglises in the Ariège valley, central Pyrenees.

carved animals, abstract signs, points and lines. Alex Hooper told me:

> "One of the most curious aspects of Les Eglises is why the artists chose such an awkward place to carry out their work. There are many surfaces elsewhere in the cave which would have been more convenient for decoration, and there are also several other narrow galleries just like the painted one which seem to have been ignored (except for some tiny spots of red ochre here and there.)"[1]

To carry out their experiments, Hooper and Collison installed an electric cable nearly a quarter of a mile long to power a series of floodlights and an ultraviolet lamp. Their intention was to study the response of the cave art to wavelengths of light beyond the ordinary visible range, with the aid of the lamps, special filters and infrared film. They hoped to detect normally imperceptible differences in pigments from one painting to the next, and to see if the sketchy and schematic outlines of animals had once been more complete. Alex Hooper explained further:

> The negative answers to our questions are none the less interesting. It seems as if the paintings at Les Eglises are incomplete not because they've deteriorated badly, but because they were

A lively black painting of an ibex from the Salon Noir in the cave of Niaux, central Pyrenees.

all executed in the same rather schematic style. And since we couldn't detect changes in the red pigment used for the paintings, it does seem probable that they were all created at the same time.[2]

While the published conclusions of Hooper and Collison are cautious and emphasize many possible alternatives, it is easy to imagine that one artist sought out this obscure nook and decorated it rapidly on a single occasion with a wide variety of animal and abstract designs. The existence of private Paleolithic art, executed for personal satisfaction or for magic motives in places hidden away from everyday view, does appear probable.

Just as surely as these intimate contexts imply a personal art, there also exist cave settings on such a grand scale that public and ceremonious uses of the chamber at once spring to mind. Of all the examples that could be mentioned, perhaps none excites the casual visitor more than does the famous Salon Noir inside the vast cave complex of Niaux in the Pyrenees. Just opposite the entrance to Les Eglises in the Ariège valley rises a great limestone spur over a thousand meters high, its interior riddled and scoured out by powerful streams issuing from the flanks of glaciers which filled the valley during the coldest phases of the Ice Age. The walls of the cavern are in many places polished smooth by the abrasion of fine sand which now lies in great heaps in the more spacious chambers. The modern visitor pads silently along unlit corridors for a half hour before the cave roof begins to rise beyond the reach of torchlight and the path turns into a steeply mounting dune. Suddenly one arrives in a circular hall, the Salon Noir, nearly a kilometer from daylight, where the

Black bison in the Salon Noir with two red and two black arrows crudely painted on its side.

roof is lost in darkness and distinct panels and niches have been worn smooth in the walls.

At this point the sense of expectancy in the visitor is strong, encouraged by the guides, who extinguish all their lights on arrival in the Black Hall. Without warning, a guide abruptly flashes his light across one of the panels, and a magnificent black ibex springs to life, the outline as taut and sharp as if it were painted an hour before. The guide illuminates each panel one by one, revealing great beasts (the largest nearly a meter and a half long) charged with vigor and personality. The effect is so dramatic that one has no difficulty in imagining some such ritual enacted in prehistoric times.

The evidence for a ceremonious use of the images is there in the form of arrow shapes crudely painted across a few of the beasts in a variety of pigments, though the postures of the animals themselves are lively and show no signs of wounding. In the echoing spaces of this great hall, the imagination readily creates the actors in some sacred piece of Magdalenian mime, who perhaps carried branches of the vegetation which is so remarkably absent from realistic cave-art scenes, and magically wounded the painted beasts in the course of the drama. But if our response to cave art is not to lapse completely into such daydreams, the hard evidence for Paleolithic activities inside the painted caves must be fully considered.

The Caverns of the Volp

Until recently this hard evidence was based almost entirely on three caves in the foothills of the Pyrenees—Montespan, Le Tuc d'Audobert, and Les Trois Frères (the last two form separate parts of the same cave system, and the River Volp runs through both of them). The three caverns were all explored just before or after the First World War, and the exceptional remains preserved in them influenced a whole generation of prehistorians in their views on the meaning of cave art. Far-reaching conclusions were drawn on the basis of very inadequate descriptions of these caves which were partly due to the physical difficulties of recording the remains accurately many hundreds of meters underground. So, although both the Volp caverns were first actively explored by the Bégouën family before the First World War, a coherent account of the remains was published only in 1958, while an actual plan of the most impressive of these remains —the pair of bison sculpted in clay in the deepest chamber of Le Tuc d'Audobert—first appeared in 1977, together with many fresh observations.

The exploration of Le Tuc and Les Trois Frères by the three Bégouën boys is the most romantic of all stories of cave-art discovery. The natural obstacles which they and later visitors had to overcome inspired evocative descriptions which did little to clarify the precise nature of the remains there. Perhaps few visitors would keep a level head after traversing an underground river by boat for eighty

Count Bégouën and his three sons, photographed at the entrance to Le Tuc d'Audobert in 1912, shortly after the discovery of the clay bison.

meters while lying flat to avoid contact with the ceiling a few inches above their head. After disembarking at an underground beach with traces of occupation and carving on the walls, the visitor arrives in a hall full of dazzling white concretions. At one end of this hall is an almost vertical ten-meter chimney, up which the Bégouën boys originally scrambled, but which is now scaled by a ladder. Then one reaches the spot where, in October 1912, Max Bégouën smashed a stalactitic drapery sealing a passage only half a meter wide leading to a succession of further galleries and chambers which had remained intact since Paleolithic times. At various points along the route prehistoric footprints and (in the narrow sections) the marks of hands and other parts of the body are visible, always overlying traces of cave bears which once had made their dens in these galleries. There are bear skulls which have been deliberately smashed for extraction of the teeth (a necklace of perforated teeth lies abandoned on the ground) and flint tools here and there, but no sign of any intensive occupation. Finally one arrives at a cul-de-sac seven hundred meters from the entrance where, modeled up against a big block detached from the ceiling, are the famous clay bisons. At this point in the visit, one German prehistorian, Herbert Kuhn, was overcome with emotion:

> We could not suppress a cry. The Bison Sculptures. Each beast is about two feet long as we see it by the flickering light of our lamps. These sculptures are no primitive carvings, they are works of surprising plastic beauty, filled with an astounding force of expression and rich with indwelling life. The nearer of the two animals is a cow and behind her stands the bull just about to mount. Obviously a piece of fertility magic. . . .[3]

Kuhn's sexual interpretation may be correct, but it is overconfidently stated. The nearer of the two bisons is carved with a clear vulva, while the bison behind, sculpted on a slightly different axis, seems more massive because of the larger boss on its back but has no sex indicated.

Furthermore, the pair of bison are not isolated sculptures, for there were other modeled figures close by which have been seriously discussed only in the study published in 1977. A third bison, which is scarcely more than a silhouette cut in clay and is only a fifth the size of the celebrated pair, was found lying close to the opposite side of the block; this was removed for safe keeping in 1912 and is now stored in a Paris museum. This little bison can still be stood upright today on the clay base which fills the gap between the front and rear legs. A tiny abstract carving (could it represent the mane of a horse or a bison?) was noticed in 1976 on the surface of the main block close to where the little bison was found. Opposite it there is a fourth

The clay bison in Le Tuc d'Audobert. The plan published in 1977 is shown below.

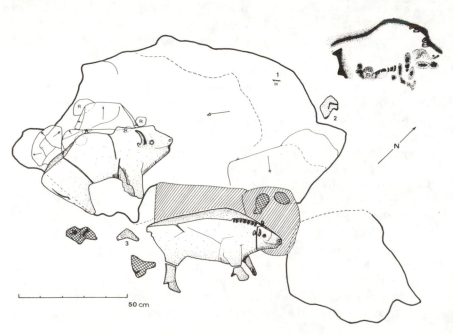

bison, approximately the same size as the first two, but merely out-lined with rather crude strokes in the clay floor. Until recently this was accepted as an abandoned preliminary sketch for a sculptured bison, but in the recent study the amount of detail present in the design and its positioning in the same rough alignment as the other figures is taken to show that it may have been intended as a com-pleted, though badly finished, relief carving in the floor. In the earlier account written by Breuil and Bégouën, there is mention of at least one heap of clay visible in a small chamber some twenty-five meters away, which they suggest could be the remains of a sculpture decomposed because of a higher level of humidity. The relationship of the four bison which have survived, grouped around the same block, is surely not fortuitous, but we may wonder if something more complex than "a piece of fertility magic" was involved.

In any case the main evidence for Paleolithic activities at Le Tuc comes not from the immediate vicinity of the bisons themselves but from twenty-five meters away, on the smooth floor of the small low-roofed chamber which opens off the gallery of the bisons at a slightly lower level. Besides the heap of clay which may be a decom-posed sculpture, there is a hollow which looks as if it could have been the source of material for one of the main figures. A fragment of stalactite apparently used to cut out the block still lies perched in an inclined position on the edge of the hole. Five crude rolled-up "sau-sages" of clay, perhaps useful for modeling details of the sculptures, lie on the floor in a corner of the chamber. There are some extremely simple abstract finger tracings, together with about fifty deep heel-marks visible in the clay and coated with a wafer-thin layer of calcite. These prints run in five or six lines, diverging in a fan toward the entry to the gallery of the bisons.

What was going on in this chamber? One step across the fragile coating would instantly destroy the impressions, so that no one can study these remains at close quarters. A French cave-art scholar, who was one of the few people to be conducted around Le Tuc in recent years, told me that he wished he had taken binoculars in order to study the floor more closely. The absence of any toeprints suggests that whoever walked across the chamber did so in a peculiar fashion. The fact that the ceiling is at no point higher than one and a half meters, combined with the small size of the deeply impressed heel-marks, convinced Count Bégouën that the chamber had been used for the ritual initiation of adolescents; so he declared that,

> since the heelmarks clearly follow five different paths, there is every right to think that this march was directed and ordered, indicating a sort of ritual dance. The fact that the heels are those of young men between thirteen or fourteen years at the time of puberty, and that five sausages of clay in a phallic shape were

dropped in a corner, awakens the idea of initiation ceremonies so frequent among primitive peoples.[4]

From assumptions of this kind it was a short step to imagine other parts of Le Tuc as the settings for elaborate rituals as, for example, Max Bégouën did in his novel *The Clay Bisons,* published in 1925:

> The Shaman and the neophytes rejoin the initiated who are awaiting them. In the little low room which they have just left, the deep and clear heelprints cross over the ritual signs which the Shaman has carved in the earth. . . . The sacred dance then begins to the beat of the magic drum. With bodies covered in long furry skins and heads hidden in horned masks, the men turn around the altar where the two clay bisons symbolise the fertility of the herd. In the yellow light of the lamps, a vapour from all the men's perspiring bodies is seen to rise. . . .[5]

The idea of dancers clad in animal skins and masks, which features in many other popular accounts of prehistoric art, was based largely on another exceptional discovery in the innermost recesses of the Volp caverns, the so-called "Sorcerer" of Les Trois Frères. This strange carved and painted figure features parts of animals such as cervid ears and horns and a horselike tail, joined to human or semihuman parts such as staring eyes, a pointed beard, and genitals (though these are placed in an odd position). The attention which the figure commands, however, is due not merely to its unusual hybrid character but also to its positioning on all fours in a chimney four meters above the ground in one of the most intensively decorated cave sanctuaries ever discovered. The Abbé Breuil spent ten months, spread over the course of ten summers, in Les Trois Frères, much of that time devoted to the arduous tracing of the hundreds of carved animals intertwined over the convoluted niches and corners of the Sanctuary, some of which can be viewed only from positions of considerable discomfort.

High above this chaotic engraved bestiary, the Sorcerer stares down like a presiding spirit from his seemingly inaccessible chimney (which can, in fact, be scaled by a series of acrobatic contortions). He is the only carved image in the chamber to be accentuated with black-painted outlines. To Breuil and Bégouën, this creature was at first taken to be an actual representation of a Paleolithic sorcerer or magician clothed in a ritual costume. Later they revised their opinion and made the more reasonable suggestion that it depicted a spirit or deity "ruling the multiplication of game and the hunting expeditions." However, many popular writers preferred the initial, more colorful suggestion of a disguised magician, and this was reinforced

The "Sorcerer" of Les Trois Frères, based on the Abbé Breuil's famous copy. The angle of the Sorcerer is actually that shown here, on all fours, not the vertical pose in which it is reproduced in many books.

by the reproduction of Breuil's tracing of the Sorcerer in innumerable books on prehistory.

We cannot be absolutely certain that the Paleolithic explorers of Les Trois Frères and Le Tuc were faced with exactly the same difficulties and obstacles that bedevil modern visitors. For example, at least four separate entries to the galleries of Les Trois Frères are presently known, and it is possible that a tunnel now blocked with clay once provided direct access between the clay bisons of Le Tuc and the Sanctuary of Les Trois Frères, which probably lie less than a hundred meters apart. However, the only Paleolithic occupation deposits (solely of the middle Magdalenian period) occur at the extreme end of each cavern, near the "beach" of Le Tuc and the Enlène entry to Les Trois Frères. This fact, along with the relative scarcity of footprints in many areas, suggests that prehistoric expeditions to the innermost sanctuaries were as arduous and rare as they are today. (Le Tuc is frequently inaccessible whenever heavy rain raises the level of the Volp at the low entrance.) That both children as well as adults penetrated into these deep and tortuous galleries seems certain, but the possibility of regular ceremonies on the scale imagined in *The Clay Bisons* seems improbable.

An imaginary cave ceremony drawn by Breuil for a children's book, combining the Montespan clay bear with "Sorcerers" dressed in skins. A delightful expression of the hunting magic theory.

A photograph of the clay bear at Montespan.

The Rites of the Cave Painters

Such exceptional remains as sculptures and footprints are likely to have survived only in very remote chambers where they are to some extent protected against casual explorers, and where conditions of temperature and humidity remain fairly constant regardless of weather changes in the outside world. At Montespan, the third Pyrenean cavern to provide prehistorians with seemingly incontrovertible evidence of magic ceremonies, clay figures and footprints were again preserved, this time guarded by even more formidable natural defenses than exist at the Volp caves.

For over two kilometers the stream known as the Houantou near the village of Montespan flows underground through a series of galleries which require the visitor to wade in a crouching position for much of the time in order to avoid impact with the low and irregular ceiling. In certain conditions of rainfall a siphon two hundred meters from the dangerous entrance at Montespan is entirely filled with water. Through this siphon Norbert Casteret swam with great daring in 1923, and emerged to discover a gallery filled with Paleolithic engravings and the remains of sculptured animals. As at Le Tuc, there were figures sketched in the clay floor as well as modeled up against the wall of the gallery. The frieze of sculptures was on a much larger scale than the bisons at Le Tuc, but seems to have been more crudely worked and is less well preserved; only two legs of a lion or horse remain in position.

A little further down the gallery, however, was a free-standing mass of clay nearly two meters long, shaped like a headless sphinx, the rounded rear quarters and remains of forepaws seeming to indicate a bear. Somewhere in this low-roofed gallery Casteret picked up a bear's skull, which Breuil and Bégouën subsequently suggested had been attached to the headless sculpture. After visiting the gallery themselves not long after Casteret's initial exploration, they also claimed to have found a hole in the bear's neck suitable for a peg presumably supporting a real bear's head. They also claimed that the bear sculpture had been polished smooth by frequent contact with some such surface as an animal skin, and that the surfaces of both the bear and the legs of the feline or horse were heavily punctuated as if spears had been aimed at them; Bégouën said that the bear was "literally riddled" with such holes.

This evidence pointed strongly to the type of hunting ritual or drama which in the minds of many commentators seemed a logical explanation for the remains that had been found in the Volp caverns a decade previously. However, just as the presence of two other bisons beside the sculptures at Le Tuc may indicate something more complex than fertility rites, so the evidence from Montespan is not nearly so unequivocal as early commentators assumed. In 1950 Paolo Graziosi, the Italian prehistorian, visited the gallery and claimed that the hole in the bear's neck was no different from the other holes in the sculpture, which in turn seemed identical to naturally occurring marks in the clay floor of the chamber. The surface of the bear was not polished but, on the contrary, rather rough. The low roof of the chamber at this point, not to mention the dangerous passages elsewhere in the cave, suggest again that whatever ceremonies were enacted at Montespan are unlikely to have been an everyday occurrence or to have involved many people.

With very few exceptions, signs of intensive occupation do not occur in the depths of painted caves—the hearths found far inside the Pyrenean cave of Labastide are one exception. So we can imagine that Paleolithic people were as disinclined to stay underground for long periods as archaeologists are today, for reasons of cold, dampness, darkness, and lack of ventilation. For these reasons systematic excavations of painted chambers are rare or very limited in extent; the French archaeologist Annette Laming-Emperaire was faced with a strike of her labor force because of unpleasant conditions on the second day of digging in the engraved galleries of the Horse Cave at Arcy-sur-Cure.

However, the execution of the finest polychrome paintings, like the huge frieze of horses and cervids at Tito Bustillo in Cantabria, must have consumed considerable time and effort. At Tito Bustillo the patience of archaeologists was rewarded by the unearthing of the Magdalenian artists' studio, an intact deposit lying beneath the magnificent painted ceiling. The excavation revealed flint tools, bones, engraved spears, and pieces of coloring material sometimes still clinging to the barnacle shells in which they had been mixed. The dig also permitted the archaeomagnetic dating of Tito Bustillo to 9,200–9,400 B.C. There are many less spacious and congenial decorated chambers than this one at Tito Bustillo, and many cases where the original Paleolithic floor has long since been washed away or buried beneath thick stalagmite (for example, in the Sanctuary at Les Trois Frères). To some extent, then, the inactivity of archaeologists inside painted caves is understandable; far less excusable is the negligence with which remains simply lying on the surface of floors inside the caves have been recorded.

A classic instance is Lascaux where, seven years after its discovery in 1940, pieces of ocher could still be picked up from the ground surface. By that time hundreds of visitors had already been conducted around the entrance hall to wonder at the extraordinary freshness of the color and the vitality of the great black bulls. The visitors trod on a concrete staircase and pathway which had been laid down directly over a calcite floor, thinly covering original Paleolithic occupation material (the Abbé Breuil had found traces of conifer charcoal and flint and bone tools). No systematic excavations of this area nor of the rest of the cave were undertaken either before or after the public opening, with the exception of some limited soundings at the extreme ends of the two long decorated galleries. In one of these extremities, a hearth with a few horse bones was discovered, and close by there was a piece of burnt rope made of plaited fiber less than a centimeter thick, which left a clear impression in the clay and is still the only evidence of Paleolithic ropemaking.

The most interesting evidence for unusual activities in Lascaux comes from a naturally formed shaft which opens in the floor about thirty meters from the entrance hall. The shaft is about five meters deep, and at its base opens out into two long fissures extending into the rock. Several facts suggest that this pit was the focus of ritualistic practices in the cave. Firstly, the pit is located at one end of a small chamber with a dome-shaped ceiling, its surface covered with scores of engraved animals and signs often obscured by a mass of confused scratches and scribbles, and mostly located far beyond the reach of an artist standing on the present floor surface. The atmosphere of this chamber reminds the modern visitor of a small vaulted chapel in a cathedral, and perhaps the setting awoke special feelings in the Ice Age engravers, too, when they covered the remotest niches with labyrinths of carved lines. The most unusual abstract signs, some of them painted, are clustered on a "cupola" over the mouth of the shaft, so high up that this part of the ceiling could have been reached only with the aid of ladders. The lip of the shaft below is polished and blackened from the movement of people against it, presumably as they swung themselves over the edge down a rope to the bottom.

Here, on one side of the small chamber at the foot of the shaft, opposite a painted horse's head, is the most celebrated and puzzling composition in cave art. No two experts have agreed on the relationship between the various participants in the scene: the disemboweled bison, the birdheaded phallic man falling backward, the bird on a stick, and the rhinoceros with six dots under its tail. It has been variously suggested that there was no connection intended between certain of the figures, that all of them are involved in a simultaneous action or scene, or that they form consecutive parts of a single story line.

In any case one reasonable assumption, that the composition recorded the actual death of a bison hunter, prompted Breuil and a collaborator to excavate the clay and sand layers at the bottom of the shaft in the hope of uncovering a burial. Instead, the excavations revealed a large number of small stones, each of them having one slightly hollowed and blackened surface, which had clearly served as handlamps. A few isolated lamps had been picked up in scattered parts of the cave by the local schoolteacher at the time of Lascaux's discovery, but the presence of so many lamps concentrated in the shaft is an odd fact. Were they deposited there as offerings, or as the refuse of some ceremony? Together with the lamps was a large quantity of conifer charcoal (which yielded Lascaux's radiocarbon

The scene painted in the shaft at Lascaux.

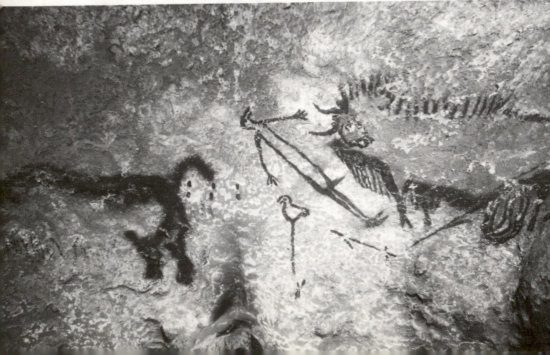

date of about 15,000 B.C.), traces of flint-working, and a number of superb highly polished antler spears. The presence of these objects in a part of the cave which cannot have been occupied, beneath a painted composition of such an extraordinary character, seems to hint at the performance of rites, evidence for which is singularly rare at many caves of lesser importance.

We can, of course, only guess at the nature of these rites and their link with the actual execution of the paintings and engravings, but recent work by Lorblanchet in an obscure cave in the arid pasture-lands of Quercy suggests how peculiar they may have been. The cramped entrance of Les Escabasses is filled almost to the ceiling with a great mass of tumbled stones and refuse four or five meters high which accumulated over thousands of years of human occupation stretching from the Paleolithic period right down to the early Iron Age. Scrambling awkwardly over the top of this deposit, one descends into an unprepossessing cavern which soon narrows to a sinuous corridor covered with black, sooty graffiti traced by the lamps of modern visitors. Underlying these ugly inscriptions, the walls are covered in a dull, bubbly coating of white calcite for much of their length. A recent geological study has shown that this layer is the result of a natural process of corrosion still going on and that began before the traces of red and black paintings in the cave, but subsequently destroyed most of the wall surfaces that existed in Paleolithic times. So at Les Escabasses we find only brief stretches of the smooth ancient brown limestone, decorated here and there with paintings so faded and insignificant that a casual visitor might easily explore over three hundred meters of gallery without noticing anything at all.

About a quarter of the way along the narrow corridor, opposite an extremely faint painting of an ibex, the guide invites one to slide on one's back across the floor and under a shelf of rock which projects out over the pathway. This exercise is comfortable only for the slim, because the gap under the overhang is a mere forty to fifty centimeters high. Peering at the rock surface just beyond one's nose, it is possible to make out three cometlike splashes of dark brown paint running across each other in different directions. The texture of the paint was evidently very fluid and probably composed of watery clay. Somebody, then, crawled under this narrow overhang and either blew mouthfuls or flicked a brush of some sort dipped in this substance (although there is nothing to prove that this took place in Paleolithic times). Other peculiar traces of dark paint appear elsewhere in the cave, including one faded round splash a few centimeters wide which seems to have hit the wall next to the painted ibex with some force.

Evidence of spattering of paint at other caves is rare, but not entirely absent. For example, one of the panels in the central chamber of La Peña de Candamo in the Asturian province of northern Spain is covered with splashes of red paint up to the ceiling (a height of eight meters), on top of which animal paintings were later executed.

More common features of the Cantabrian caves are inconspicuous streaks of charcoal (these also appear in Les Escabasses). The little black lines, which are often found on protruding corners or on stalactites, are too simple to be called abstract signs, but are too definite to be dismissed as accidental torch marks. It may be that markings of this insignificant kind, which have often been neglected in published studies of particular caves, are related to a general idea of some kind of rite, such as the magical treatment of surfaces before decoration.

If this is true, then it could help to explain the anarchic scribbling of carved lines which creates such havoc among the beautiful animal engravings above the shaft at Lascaux, or at Altxerri (though in this Basque country cave the wild lines sometimes follow the outline of a painting and add to the modeling of the figures). It seems likely that there was a need to prepare or conceal surfaces in some caves for reasons which are difficult to imagine.

Further progress in reconstructing the rituals of the cave painters or in establishing who took part in them is unlikely without a rare combination of circumstances. The first essential is the preservation of original Paleolithic ground surfaces and deposits intact, despite the action of thousands of years of erosion and concretion. No less vital is the diligence of potholers in reporting their discoveries without wiping out the evidence by, say, a single imprudent march across a clay floor. Two recent finds more or less fulfilled these conditions, and when the work of specialists at both sites is over, their evidence may add many intriguing new details to our picture of the activities of the cave artists.

A panel of black paintings in the cave of Les Escabasses. Intensive study suggests that *(left to right)* a small abstract sign, a headless horse and another horse were all painted with the same pigment. Far right, close-up of a splash of dark paint that hit the cave wall with some force.

The Underground World

The first of these discoveries takes us back to December 1970, when speleologists over one kilometer inside the mountain of Niaux finally succeeded in pumping dry the last of a series of three subterranean lakes. This was the climax to nearly fifty years of sporadic efforts to penetrate the farthest reaches of the cavern despite the blocking of galleries by seemingly impassable water traps.

The first of these, the "Green Lake" which lies almost half a kilometer beyond the black beasts of the Salon Noir, was crossed on a raft during a dry spell in the winter of 1925. On the walls of the short gallery beyond, traces of more black paintings were discovered, but another siphon barred further exploration. The second siphon was crossed only with difficulty in 1949, but the third and final lake defied the most courageous efforts of potholers until pumping equipment was installed in the winter of 1970. Then, at long last, a team was able to penetrate into a further kilometer of previously unknown galleries, the whole length of which was impatiently explored by the discoverers, although prehistoric footprints were noticed on sandbanks in two places. Nor did the speleologists bother to inform local archaeologists until January 1971, by which time a television broadcast had actually been made from the new gallery. Thereafter it was sealed naturally by heavy rain and later by a wall constructed under the direction of archaeologists to control the access and environment of the gallery.

After the initial studies of the specialists had been completed, equipment designed to monitor changes in humidity and temperature was installed, connected by nearly two kilometers of cable to a research cabin built near the modern entrance to Niaux. The sensitive monitoring devices are at present continuing to provide scientists with the first record of how an intact painted cave responds to daily and seasonal weather changes outside. When this information is compared with similar data from the Salon Noir, which is heavily visited by tourists, it may eventually throw light on the problems of conserving paintings and engravings far underground.

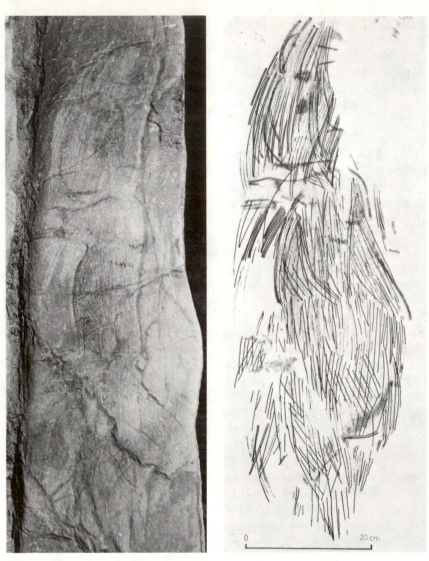

The cave of Altxerri in northern Spain contains several bison depicted at an extraordinary vertical angle. Many of the artworks are overlaid by scribbled lines which in some places obscure and in others emphasize the animal outline.

For the time being, the chief interests of the new gallery at Niaux are some five hundred well-preserved human footprints, and their relationship to five black animal paintings grouped in one part of the gallery. These paintings, which seem to be rapid sketches unlike the individualized and detailed animal portraits of the Salon Noir, include one horse, three bison, and one of the very few depictions of a mustelid (probably a stone marten or a weasel) known in the whole of Paleolithic art. The uniqueness of this image corresponds to a coincidence of an equally unlikely order. While studying the short gallery between the second and third siphons, the archaeologists came across the intact skeleton of a stone marten, and at once suspected some magic practice originally involving the body of the stone marten with its depiction on the cave wall. However, a close study of the animal's tracks showed that it had fallen into the cave through some natural fissure long after the execution of the paintings and had desperately sought a way out of the gallery until it reached the second lake, where it died of exhaustion. The animal tracks everywhere overlie the footprints of human visitors (among them at least three young children) who dropped well-preserved torch ends in various parts of the gallery.

The carbon from most of these torches has yielded dates centering around 8,000 B.C., long after the end of the Magdalenian period when cave art ceased (an even later visit during the Neolithic period also seems likely). As for the torches and prints of the artists themselves, these seem unfortunately to have been washed away from the sandbank immediately in front of the paintings. However, a variety of evidence, such as the absence of remains of any period between the second and final lakes, does suggest that an unknown entry to the new gallery was used in prehistoric times, and that the incredible obstacles faced by modern explorers were not penetrated in the past. While the new information from Niaux is disappointing in its direct relevance to the activities of painters in the gallery, it does show that barefoot visits to muddy caves by young children as well as adults continued long after the end of the Ice Age.

The second discovery, which occurred just over a year after the final lake was pumped dry at Niaux, was more spectacular and more unexpected. The obvious possibility of further galleries beyond the Niaux lakes had tantalized potholers for decades, but at Fontanet (not far from Niaux in the Ariège valley) the galleries had been used as a training ground for generations and it was assumed that every inch of them had been explored by enthusiasts. It was in February 1972, however, that one such band led by Luc Wahl was returning from the depths of Fontanet toward the entrance, keeping an eye on the roof of the corridor as they walked along. Not far from daylight (at a stage where most previous explorers were either filled with thoughts of home or intent on the dangers to come), Wahl noticed a small opening in the ceiling some five meters above. The team immediately climbed up to it with the aid of a nylon ladder. Together they all walked through over three hundred meters of gallery until, in the spacious chamber near the end, their torchlight fell on a vivid red-and-black painting of a bison. With immense care Wahl and his colleagues threaded their way back exactly along their own footprints and at once alerted the authorities.

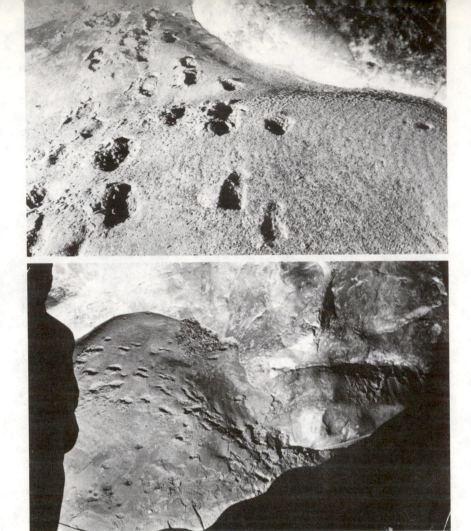

Views of the footprints of three children who walked across a sandbank in the decorated gallery of the Réseau René Clastres, Niaux, around 8,000 B.C., long after the paintings had been executed. The gallery was discovered in 1970.

Today Fontanet is one of the most important and best guarded of all the painted caves. With considerable ingenuity Luc Wahl, now the official guardian of the cave, has constructed a series of impregnable gates and a system of aluminum catwalks suspended above the ground on nylon ropes fastened to the cave walls which permit study of the floor below without risk of destroying the evidence.

That evidence is indeed remarkable. In the first part of the cave there are no traces of decoration on the walls, but the clay floor surface underneath the catwalks is covered with foot and hand impressions in an almost perfect state of preservation. If one lies on the

catwalk and shines a torch at an angle across the handprints, the folds
of the skin can be distinguished and, in one case where a knuckle had
been planted momentarily in the floor, the shape and length of the
fingernails. Here at Fontanet, just as in every known case of footprints preserved in Paleolithic caves, the feet were unshod and many
of the impressions clearly belong to young children. The other remains in this part of the cave are less spectacular. In two places crude
finger drawings have been made by stabbing at the clay in circles and
lines, without any obvious representational meaning. There are also
two flint blade tools of a typical Magdalenian form lying on the
ground exactly where they were originally dropped.

Beyond this first section of Fontanet the gallery becomes narrower, and although the modern visitor passes easily along the metal
catwalks, the original path was by no means as convenient. In one
place there are blurred impressions of fingers as they tried to grip
the slippery sides of the clay for support. As the gallery widens
slightly again, there are the first traces of occupation—the remains
of a small hearth, and the intact skeleton of a salmon lying at shoulder
height in a rocky cleft where it had presumably been casually tossed
at the end of a meal.

Soon after, one arrives in the last chamber of Fontanet, the only
place where the natural topography of the cave becomes at all impressive. The passage progressively widens into a spacious hall with
a high roof, sealed at one end by a dark mass of tumbled scree,
suggesting that perhaps the outside world is not very far away. But
the attention does not linger for long on these natural features. Here
in the middle of the hall there are no footprints preserved, but the
surface of the soil is black with a great mass of charcoal and burnt
bone. Here at one's feet lies the debris of the last Magdalenian feast,
with its litter of jawbones and shoulder blades and the occasional flint
tool. Immediately above the hearths are the paintings and engravings. At first, close to a niche on the right-hand wall, one notices two
grotesque human heads with bulbous noses. On the adjacent side of
the niche is the black outline of a bison and one headless anthropomorphic figure. None of these paintings are particularly impressive, but if a torch is held close to the rock surface beneath them,
a tangle of fine-line engravings suddenly springs to life, including
several magnificent, vigorous bison heads, their bodies "pierced" in
places by little triangle-headed arrows. A little farther along from the
niche are two small ibexes or Pyrenean izards, which very closely
resemble those in the narrow painted cul-de-sac at Les Eglises a little
farther downstream in the same valley.

The most striking of the artworks at Fontanet, however, is the
red-and-black bison which first caught the attention of the original
explorers, standing isolated on the opposite wall of the cave, glower-

ing over the chaos of bone and charcoal on the floor below. This beast conveys an impression quite unlike the impassive and idealized quality of the bison from the Altamira ceiling or the Salon Noir at Niaux. Its ferocious spirit is accentuated by the conspicuous genitals and by the freshness of the ocher, which in places runs in blood-red drips from its stomach and flanks.

To stand silently in this final chamber at Fontanet is one of the strangest experiences that prehistory can offer. One hundred and fifty centuries separate the visitor from the cries of the children who slipped their way around these muddy galleries, from the casual tossing of a salmon carcass over a shoulder, or from the instant when the red paint dribbled over the black outlines of the bison. Since then, nothing has happened to disturb the footprints, the charcoal and the bones, which resemble a waxworks exhibit of prehistoric man rather than an archaeological deposit. In the ordinary conditions of digging, one sometimes encounters hearths and tools decomposed almost beyond recovery, or else fused into almost solid rock by the processes of concretion, but here everything lies exactly as the Magdalenians saw it when they finally deserted their living place and sanctuary. One feels their presence so intimately that there is an absurd temptation to look over one's shoulder to be sure that no one else is standing in the chamber. However often the thoughts of the Paleolithic artists may seem odd and remote to us, there are places in the painted caves where they seem oppressively near to our own irrational imaginings, encouraged by the strange shapes and echoing noises of the underground world.

Children's handprints preserved in the clay at Fontanet.

10: The Elusive Meaning

"A True Passion for Art"

The two discoveries at Niaux and Fontanet, occurring little more than a year apart in 1970 and 1972, surprised prehistorians for at least two reasons. Firstly, at Fontanet an extensive occupation deposit lies directly beneath decorated cave walls which are today situated far underground, although originally they may well have stood close to an entrance that is now blocked. Even supposing that the fierce red bison and the other paintings and engravings once stood in the half-light from an open porch not far away, the association with the feast remains is unusual. It warns us, perhaps, against assuming that all painted chambers were holy and untouchable places simply because so many caves have a cathedral-like atmosphere today. Secondly, the drawing of a species previously unrecorded in Paleolithic cave art, the mustelid in the new gallery at Niaux, further suggests that any rigid rules or explanations which we try to make are likely to be continually enlarged and modified by new discoveries. The variety of settings so far described—from the soaring vault of the Salon Noir to the cramped cul-de-sac of Les Eglises—surely also suggests a variety of motives on the part of the artists. How far, then, is it possible or even worthwhile to pursue "the meaning" of Paleolithic art?

The broadest argument of all, and it was also the earliest to be advanced by prehistorians, is that the Ice Age hunters were simply enjoying themselves, indulging in animal painting and engraving for no other reason than personal creative satisfaction. This idea arose among nineteenth-century archaeologists long before any painted caves had been discovered, at a time when the very existence of engravings of extinct animals on bone tools dug up from so remote an epoch aroused wonder and admiration. From the beginning Lartet and Christy, who first drew the attention of the world to Paleolithic art, were less concerned with explaining the designs than with justifying the existence of the artists. Their investigations of the massive deposits full of animal bones near Les Eyzies convinced them that life in the Ice Age was a relatively congenial and relaxed affair. As they wrote in a French journal in 1864, just a year after their first excavations:

We should remember that hunting and fishing amply furnished the needs of these aborigines, providing them with the leisure

197

An Ice Age art studio: This engraving appeared in 1870, influenced by the early "art for art's sake" theory concerning the meaning of the engraved and sculptured objects.

of a fairly untroubled existence. Thus, if necessity is the mother of invention, we could say that the leisure of an easy life gave birth to the arts.[1]

Many subsequent commentators uncritically supported the idea that the engravers were merely following an instinctive taste for artistic expression:

> neither a religious nor any other outside purpose can be proved to be intended in the pictorial works of primitive men. We have a full right, therefore, to give credit to the numerous witnesses who assure us that these representations originate from the pure pleasure of making them. . . .[2]

Such a straightforward solution distracted attention from the actual content of the carvings, or at best encouraged detailed explanations of an absurdly facile kind. So a composition carved on a bone from La Madeleine which had baffled Lartet was interpreted by the British archaeologist Boyd-Dawkins as a realistic hunting scene. He did not try to explain why the horse heads were drawn at a different scale from the curious human figure with a bulbous head, nor why that figure was shown naked when, as Boyd-Dawkins admitted, hunters must have been clad in skins. "He has evidently surprised a herd," Boyd-Dawkins wrote in 1880, "and the head of the horse which he is attacking has its ears pricked up in a very significant fashion."[3]

One of the only commentators to express any surprise or bewilderment at the subjects and scenes of mobile art was the leading French prehistorian, Émile Cartailhac, who in 1876 found himself at a loss to account for the engravings executed at Laugerie Basse near Les Eyzies:

> Here there is something other than the proof of a marvellous artistic disposition; there are unknown motives and aims at work; these engravings which grow more numerous with every day and which were applied to decorate finished objects as well as scraps of bone, have a meaning which eludes us still. . . .[4]

Yet so strong was the prevailing notion of the aesthetic spirit of the cave artists that this glimmer of doubt was ignored even in the later works of Cartailhac himself. Reliable accounts of the behavior of peoples at a comparable level of technology to the Paleolithic hunters began to appear only toward the end of the century, and the complexity of primitive beliefs which these accounts revealed did not entirely escape attention. Thus, in his book *La France Préhistorique* (1889), Cartailhac pointed out that Lartet and Christy's theory of leisure giving birth to art must be wrong—disproved, for example, by the rarity of rock art among the South Sea islanders and its abundance among the materially impoverished Bushmen. But to leap one step further, to compare the actual *content* of recent primitive beliefs with the themes of prehistoric art, apparently never crossed the minds of the nineteenth-century prehistorians. Instead, Cartailhac's account of the earliest artists closed with a classic expression of the conventional view:

> They had a true passion for art, and at each occasion they began tracing their designs, sometimes abandoned or destroyed without regret, to reach their goal—that of personal satisfaction.[5]

The design on a baton found at La Madeleine, Dordogne, during the 1860's.

Detail of a panel of wild ox engraved with an unusual "cameo" effect through a thin natural black deposit at La Loja, northern Spain. Note the seemingly haphazard arrangement of oxen at different levels and scales.

Today, anyone merely glancing at reproductions of the expressive painted horses of Lascaux or Ekain could scarcely doubt that a desire for aesthetic perfection played a prominent part in the minds of the cave artists. In an interesting essay published in 1971, one French commentator, Etienne Souriau, argues that our distinctions between art and magic are meaningless in the context of prehistoric art, and that the astonishment which the modern visitor feels at encountering beautiful images in such unlikely underground settings was exactly the effect sought by the original artists. In a sense Souriau's argument represents a revival of the nineteenth-century "art for art's sake" theory. He asserts that all other possible motives which we may try to distinguish in terms of symbolism or magic are trivial compared to this artistic impact, the shock of recognition, which we still feel today:

> Everything looks as if (putting it cautiously) the artist had wanted to create a spiritual presence, as close as possible to a real presence, but not a symbolic one. . . . We should not neglect this astonishing and disquieting effect which art as magic must have evoked, independently of all ulterior aims, simply by its power to make the unreal and imaginary into a real and concrete image.[6]

The obvious problem with Souriau's arguments, and with any point of view which asserts that the intentions behind cave paintings were primarily artistic ones, is that they are based on our private emotional response. They lead us away from considering the actual content of the painted scenes, which is the only way that we can hope to test other theories and speculations.

By the turn of the century, other possibilities had become too obvious to ignore. In 1899 Spencer and Gillen published their vividly illustrated volumes on the Arunta of central Australia, which included accounts of ceremonies and sacrifices practiced with the aid of animal drawings in the sand. These rituals were directed toward the cult worship of the animal ancestor (or totem) adopted by a particular tribe, images of which were sometimes painted on remote cave walls strictly prohibited to women and uninitiated boys. Earlier, cruder accounts of totemic beliefs among the Aborigines and the American Indians had profoundly influenced the works of philosophers of religion such as Sir James Frazer, who devoted much space in *The Golden Bough* (1890) to defining what was then known of the range of beliefs in ancestral spirits. But it was specific Australian

On November 21, 1977, bush pilot and rock art expert Percy Tresize discovered this huge panel of nearly 300 paintings in a remote part of northern Queensland. Like Paleolithic cave art, many designs are superimposed at different scales in apparent disregard for previous work. Note the negative hands, which continue to be made by the Aborigines for a wide variety of purposes.

examples of remote cave art created for magical ends that caught the imagination of contemporary prehistorians. One of them, Salomon Reinach, who was head of a Paris museum, made a vital break-through in suggesting that the behavior of living hunter-gatherers was a key to the past:

> The only hope that we have to know why cave dwellers painted and sculpted is to ask the same question of living primitives.

This assertion, made in an influential article of 1903, was coupled with a decisive rejection of the aesthetic "art for art's sake" view:

> They knew what they were doing and why they were doing it; they were not dreamers or idlers, engraving or painting any familiar outline according to the inspiration of the moment. . . .[7]

The appearance of volumes of photographs and descriptions of exotic Aboriginal rituals would not, by itself, have swung opinion round so decisively in favor of magical interpretation, had it not coincided with the earliest work of a young cleric, Henri Breuil, then just twenty-six. Two years before the appearance of Reinach's article, Breuil, together with his colleagues Louis Capitan and Denis Peyrony, had discovered the faintly inscribed carvings hundreds of meters underground in the narrow twisting corridor of Les Com-barelles near Les Eyzies. A few days later Peyrony explored a neigh-boring valley and stumbled across the faded but magnificent paint-ings of Font-de-Gaume, most of them situated well beyond the reach of daylight. The existence of cave art in locations that strongly sug-gested magical motives was now beyond dispute, and several earlier discoveries which had been scorned and discredited—among them the painted bisons of Altamira—were now vindicated. At the time of Reinach's article of 1903, eight painted caves had been found, sufficient to encourage some of his bold and far-reaching assertions —for example, that Paleolithic hunters practiced "totemic cults iden-tical to those of the Australian Aruntas today."[8]

Few would deny that the animals depicted in the painted caves were the objects of some special emotion and perhaps veneration on the part of the artists. However, the argument that they were neces-sarily connected with some ancestor cult or myth has won few adher-ents, for again it is a suggestion that is hard to test against the available evidence. Had individual tribes adopted a particular animal as their totem, we might expect to find painted caves dominated by a single species, and this (with a few exceptions like Rouffignac or Covalanas) is far from being the case. For the typical situation, where cave walls are covered with a combination of several different species, we have no way of judging whether totemic stories may have inspired the designs any more than other possible myths or superstitions.

The most ingenious and bizarre theory to be advanced along these lines (although no direct discussion of totemism is involved) proposes that we may "read" the compositions of Paleolithic art as if they were family trees, symbolizing the relationships between various clans which had adopted animal ancestors. In other words, we might interpret a cave panel featuring bison and horse as a record of the marriages between the bison clan and the horse clan. As the climax to one article expounding this idea, Annette Laming-Emperaire tried to "decode" the great frieze of animals in the entrance hall at Lascaux as a detailed and complex family history; any extract will convey the flavor of this strange exercise:

> The horse which turns its back on the scene could be the ancestor of the horse clan; it has just concluded an alliance and returns to its territory. It is going to make an exchange of women with the bovid clan. The little horses which surround it and which face the bull are mares, so they are going to marry into the bovid clan: the marriage is patrilocal. . . .[9]

One of several reasons why we might find this theory wildly improbable is that recent studies of totemic beliefs among the Australian Aborigines have demonstrated that they are more diverse and flexible than was widely believed in Reinach's day. For example, the central character of the ancestor stories can be plants or rocks as well as animals. The stories do not invariably explain the descent or lineage of a particular tribe, for they often describe how certain individuals, or sexes, or cult groups, or even a group at one special occupation site, came into being. In other words, we should not expect ancestor myths to follow rigid rules, and, again, we have no way of proving that such stories are in fact present among the cave designs.

The Myth of Hunting Magic

In any case, Reinach himself did not elaborate his suggestion of totemic cults, but concentrated on an idea that was to prove far more influential. In 1903 no depictions of dangerous animals such as felines or serpents were yet known, and Reinach was struck by the artists' preference for "desirable" food animals. Although he presented little positive evidence to support his claim beyond general references to primitive beliefs, Reinach suggested that the artists were seeking to multiply the numbers of game or to control their movements by magic means, perhaps by attracting them to the vicinity of the cave. In short, the artists were preoccupied "not with pleasing, but with invoking," and the purpose of the painting and engraving was vitally connected with the survival of the hunters.

The theory of hunting magic became steadily more popular at the same time as the range of subjects depicted in the art grew with each fresh discovery. To accommodate all the new finds it was necessary to make the theory increasingly vague and even contradictory. For example, when the first pictures of dangerous carnivores were noted, this indicated that Reinach's theory of attracting beasts to the vicinity of the cave must be wrong, for what hunter would have wished to charm a cave lion in this way? If, on the other hand, it was imagined that the hunter desired to exert a practical control over both desirable and dangerous species alike, then the rarity of certain subjects (notably reindeer and small rodents) remained puzzling.

One suggestion was that the artists were only interested in luring animals that were difficult to catch, so that the docile reindeer was simply not an object of concern. However, this could not account for the paintings and carvings of birds, which are even rarer than the depictions of reindeer. As we have already seen, birds certainly contributed to the hunters' diet; the most common bones which can be identified belong mainly to small game species such as ptarmigan and grouse. What, then, can have induced the artists to portray mainly water fowl and especially ducks among the eighty or so known representations of birds?

In short, the practical application of the hunting magic theory was confusing and inconsistent, and perhaps it is surprising that this viewpoint persisted for so long. Its influence was due largely to the impact of the discoveries in the Volp caverns and of the Montespan clay bear on Breuil and Bégouën, convincing them of the importance of magical hunting rituals among the cave artists' activities. They were encouraged to interpret several debatable aspects of the art in terms of hunting magic; Breuil thought the featherlike signs on or near some animals were all weapons (though some of them look most impractical), while Bégouën saw the lines issuing from the muzzles of some beasts as blood (they could represent breath, or even the animal calling). To do Breuil justice, it must be said that he rarely concerned himself explicitly with the meaning of cave art, other than in rather vague and spiritualized terms. Instead he preferred to devote his remarkable energy and eye for detail to questions of style and to his unsurpassed copies of cave designs. Other less cautious spirits were ready to interpret certain abstract signs as pictures of traps, with which Breuil eventually disagreed, or vertical lines drawn on the bodies of a few animals (for example, at Ebbou and La Clotilde) as the ribs of flayed corpses.

Among a host of similar dubious suggestions, one plausible and coherent theory stands out head and shoulders above the rest. During the thirties an amateur prehistorian and game hunter named P. Leason became convinced that many peculiarities of style in cave art could be explained if the artists had been copying their designs from dead models. Leason pointed out the unnatural stance of many animals, which appear not to be planted firmly on their feet at all, but to be "floating" with the tips of their feet pointing downward and with a lack of tension in the shoulder and leg joints ("Many have a decidedly fairy-like poise," he wrote[10]). By careful comparison with photographs of the corpses of modern animals, Leason claimed that the study of dead animals had influenced aspects of style such as the frequent twisted appearance of horns, antlers and hooves of animals otherwise viewed in profile. In many examples, he pointed out that the artist's "sighting point was lower than the centre of the animal's body, and very often lower than the level of the feet,"[11] which would have been the case if the artist was attempting to sketch a corpse stretched out on the ground. Leason never claimed that the cave painters were intentionally representing dead animals, but simply that the study of corpses had influenced their style and had contributed to their expert eye for naturalistic detail.

Keeping this vital distinction in mind, it is clear that Leason's theory does account logically for the odd features of a large number of cave paintings—most impressively of all, perhaps, in the case of

Two of the bison on the famous painted ceiling at Altamira, which may owe their strange attitudes to the copying of dead models.

A photograph and drawing of a lion finely engraved on the surface of a stone slab from La Marche, Vienne, western France.

the Altamira bison, many of which "float" on tiptoe or are curled up as if dead. We might be tempted to go further and assume this aspect of style *did* correspond to some religious belief: the imagination easily conjures up a hunter-artist propitiating the soul of a slaughtered beast by sketching its corpse and transferring this image to a "sacred" cave wall. However, it should be remembered that though many cave animals *do* look slack and immobile, quite a number seem to be caught in extremely lively postures indeed, and it is also evident that the cave artist's eye for anatomical detail was not always perfect.

A fascinating example of this occurs among a series of felines engraved on the surfaces of small limestone slabs at La Marche, some of which, far from looking dead, appear to be poised in the middle of a ferocious assault. A few of the lion heads display an anatomical error in the positioning of the front canine teeth that is common in medieval sculpture and many popular artists' depictions, because sharp teeth protruding from the upper jaw enhance the savagery of a creature portrayed in such a manner. However, if the La Marche engraver had been faithfully copying the corpse of an actual lion, he would surely have represented the canines as they are in reality, with the lower teeth positioned in front of the upper. The La Marche lions seem to be one instance of a Paleolithic artist drawing up vigorous images purely from the imagination, rather than from studious attention to a lifeless model.

Today we are in a better position than ever before to assess the relationship between the animals depicted in cave art and those which the hunters pursued as game. In the past twenty years knowledge of the exact content of painted caves and of engraved, portable tools and objects has grown increasingly precise through the efforts of scholars, although an immense amount of truly comprehensive recording remains to be done. One key result of this work is the realization that only about 10 percent of all the animals in cave art appear to be wounded or marked with signs which look like spears or arrows. So, while some depictions *do* seem to have been the object of magical rites of destruction (like some of the black beasts of the Salon Noir), these are usually in a minority among many other untouched designs. The type of ceremony suggested by early theorists such as Reinach or Bégouën can only have applied to a small fraction of the paintings and engravings found in a typical cave.

Even more striking is the point touched on briefly above, that the animal species depicted bear little resemblance to what we know of the actual kinds of meat consumed by Ice Age people. This divergence between the economy of the hunters and the content of their art is so surprising and fundamental that it demands further exploration.

A painted horse at Ekain, northern Spain, with an engraved arrow in its flank. Such explicit "hunting magic" scenes are, however, uncommon in cave art.

The most glaring fact is that only nineteen painted caves in western Europe contain pictures of reindeer, comprising well under 150 beasts in all, the majority of them located in regions of southwest France where for thousands of years reindeer were apparently providing 90 to 98 percent of the meat content in the hunters' diet. The reindeer in fact occupies only the seventh or eighth position in the list of species selected by the cave artists, far behind the horse and bison which fill the top two places. Horse and bison were certainly hunted throughout western Europe, too, but at least in southwest France they were not often exploited in such a consistent and specialized way as reindeer. What inhibition can have deterred the artists from portraying their most important food animal?

Before jumping to any conclusions, it is important to realize how delicate the interpretation of these figures must be. It is often difficult to distinguish the outline of reindeer from those of other cervids, especially red deer, so that the figures cannot be considered totally reliable. Similarities of style have suggested to several experts that the great majority of reindeer both on the cave walls and on the small objects belong to the end of the Ice Age (the period of the middle and late Magdalenian, about 13,000 to 9,000 B.C.). However, this is also the period when the over-all numbers of occupation sites and decorated caves in southwest France seem to have increased dramatically. So, instead of some swing of taste or fashion toward the painting and carving of reindeer, it seems that in all periods their numbers remained consistently low compared to the other animals.

Are the same preoccupations, the same peculiar inhibitions, present in cave art outside the classic region of southwest France? Simply looking at the "top ten" of animals represented in the caves of northern Spain, the answer would seem to be Yes. Horse and bison again dominate the cave walls, while reindeer are rare not only in the caves (fifteen depictions in three caves) but on the tools and carved objects (three depictions found at two sites). The most common animals chosen for painting or engraving show little connection with the economic activities of the hunters. As we have seen, when specialized hunting developed toward the end of the Ice Age, the main targets of the chase seem to have been red deer and ibex. Just like their contemporaries in France, the Cantabrian cave artists did not display any special concern to depict the animals which they most frequently hunted. So, judging from these bare facts and crude figures, we would conclude that hunters throughout western Europe were decorating caves and objects with the same selection of animals and the same motives in mind, and that these motives had little to do with hunting magic.

However, the Spanish figures for mobile art conceal a surprise for anyone accepting such a comfortable view. The subject matter preferred in the mobile art differs notably from the content of cave art, much more sharply than in southwest France. There are no bison representations at all on the tools and ornamented pieces, there are twenty-eight horses, but no less than fifty-six ibexes. In fact ibexes comprise exactly a third of the realistic subjects in the mobile art, although there are only 10 percent of them in the local cave art.

There would be little point in dwelling over such tedious figures if they did not illustrate two vital considerations. Firstly, neither in mobile nor cave art is it possible to draw a consistent or convincing link with the animals that were being pursued for their meat, hides or horns. Secondly, the "rules" that governed the engravers of the bones and stones do not always seem to be identical to those of the cave artists, and perhaps these "rules" varied from region to region

and from cave to cave far more than our crude total figures can possibly suggest.

Indeed, how much importance should we attach to the frequency with which certain animals are shown, especially horse and bison? Is the apparent consistency of this art just an illusion, and how far can statistics take us into the minds of prehistoric men?

Order and Chaos in Cave Art

Undoubtedly, the figures do conceal an immense variety. We should certainly like to discover why horse and bison were so frequently chosen, but it would also be interesting to know why the engravers at just one site, Le Gabillou in the Dordogne, became infatuated with reindeer designs to the point where these take second place only to horses in the totals for that cave. According to one expert, the style of the twenty-one Le Gabillou reindeer is identical to that of the one or at most two individual reindeer visible at Lascaux not far away, solitary beasts indeed amidst a sanctuary containing more than 1,200 other animal representations. Were the artists really working to set rules, and, if so, how far can we hope to discover these rules?

In 1975 an intriguing attempt to solve this problem involved the "decoding" of the painted caves with a computer. At first sight nothing could be more incongruous than the application of a sophisticated mathematical machine to explore the intricacies of an art created tens of thousands of years before the invention of written numbers. The problem which the British computer scientist, Antony Stevens, set out to tackle was extremely simple to pose in principle, but it would have been laborious to work out in practical detail without the aid of a machine. The idea was to compare not merely the frequency of, say, horse and bison in a particular region, but to see how the complete contents of each cave compared with the next. Only by contrasting the total pattern of all the major animal species from one cave to another could we possibly hope to discover whether cave artists in one region had the same ideas and intentions as those in another.

So Stevens set his computer to print out the numbers of every important animal species in thirty-seven painted caves (other sites had to be excluded when there were too few paintings or carvings of any kind to be represented in the analysis). A study of this kind clearly has its limits. It is based on descriptions of the contents of caves published in books and papers, and many of these inventories are far from being exhaustive or even satisfactory. Perhaps we should judge the importance of a particular animal not from the number in each cave, but by their size, color, degree of complete-

Despite the rarity of reindeer as a theme of the cave artists, several magnificent treatments are known, like this engraving of the front legs, head and antlers at Altxerri in northern Spain. Notice the fox carved inside the reindeer neck.

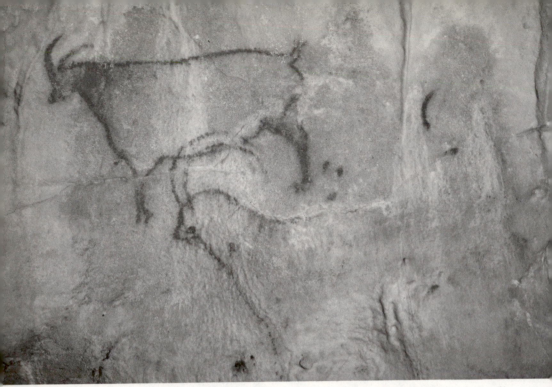

It is often difficult to decide whether animals were positioned haphazardly or deliberately by the cave artists. The obvious pairing of a male and female ibex at Cougnac is extremely rare, while the coupling of a horse and mammoth at Saulges *(opposite)* is far more ambiguous.

ness, prominence of position in the cave, and so on. Of course, to take account of all these other factors in the computer program would have been an extraordinarily complicated task. No two experts would agree where to draw the line in matters of size and positioning, whereas the simple counting of the number of animals present is likely to provoke only minor arguments.

Despite all these drawbacks, the results of the encounter between the caves and the computer are of considerable interest. The machine printed out the degree of similarity between the contents of one cave and another in the form of a chart, and from this it can be seen that two distinct groups emerge. One comprises the three caves in northern Spain with the greatest numbers of paintings and carvings (Altamira, Pasiega and Castillo), which are also geographically close to each other (indeed, Castillo and Pasiega are situated in the same mountain). In these three caves the proportions of at least four types of animals—hinds, oxen, small bovids, and stags, with perhaps bison and horses as well—are very close, suggesting that similar motives may have driven the artists in these three cases. An exactly similar conclusion also links four caves in the Pyrenean foothills (Marsoulas, Montespan, Niaux and Le Portel), where the proportions of bison and horse in each cave are much closer than we would expect by chance. It is as well not to read too much into these results. While

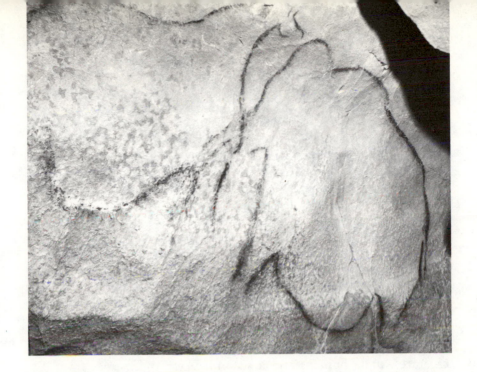

the contents of these two groups of caves coincide in an intriguing way, there are other sites in the same regions which have very different proportions of animals in them. In other words, there *is* great variety in the contents of the caves, and the counting of animals alone cannot tell us for sure whether one idea or several inspired the artists in a particular region.

Certainly the abundance of horse and bison on the cave walls would strike even a casual visitor to the caves as a notable fact. It particularly impressed the first two researchers to attempt a systematic study of the contents of the caves in the 1950s and 1960s. These two experts, Annette Laming-Emperaire and André Leroi-Gourhan, independently concluded that the prominence of horse and bison reflected not merely a special interest or veneration of both animals by prehistoric people, but a complex symbolic system of pairing present throughout cave art. Wherever horses were pictured on cave walls, it was likely that bison would be there too, deliberately associated with the horses to produce a balanced composition.

This special pairing of horse and bison was projected through all the different signs, human figures and animals which the cave artists chose to draw or paint. Works of art were not accumulated or superimposed at random, but were always deliberately positioned and associated according to a system of beliefs. So, according to the French researchers, the entire Paleolithic universe of thought was divided into two principles, represented by the two groupings of animals. Because cave art includes male and female human figures, it was natural to think of the opposed groupings in terms of male and female. Ironically, for different reasons, both workers arrived at different conclusions about the sexual identity of the principal animals depicted in the caves: Leroi-Gourhan considered that the horse represented the male principle and the bison the female, while Laming-Emperaire thought the opposite was true.

Both schemes attributed an extraordinary degree of complexity to the minds of the cave artists. Was the positioning of paintings and engravings on irregular cave niches and chambers really the product of such careful thought and organization?

The ideal layout for a Paleolithic sanctuary was determined by Leroi-Gourhan after a close study of sixty-six caves containing over 2,000 representations. He claimed that the members of his female category—such as bison, ox and mammoth, female humans and certain signs—were usually found in the center of the most prominent panels of a cave. The peripheral subjects of cave art (appearing at the edges of the main compositions, and also at the entrance and end of the cave, as well as in any side nooks) usually belonged to the male category. These included horse, stag, ibex, and other less common creatures, male humans and a second group of abstract signs. So, according to Leroi-Gourhan, the sanctuaries were organized according to definite spatial rules, and these rules helped to express the duality which he took to be the foundation of the cave artists' beliefs.

This ideal layout for a painted cave was not something merely dreamed out of Leroi-Gourhan's head, but was based on a detailed statistical study of the sixty-six caves. Subsequently, several critics were able to discredit the statistical methods with which Leroi-Gourhan attempted to prove his theory. The arguments are too complicated to be stated here, but one glaring and indeed extraordinary procedure adopted by Leroi-Gourhan deserves a mention. Just like Stevens' study, Leroi-Gourhan's work took no account of such factors as the size, color, and degree of finish of the animal depictions. Yet neither was it based on the actual numbers of each animal species. Instead, Leroi-Gourhan was concerned only with "themes" —the presence or non-presence of any given type of animal regardless of the number of individuals represented.

This procedure allowed Leroi-Gourhan to find much more order in cave art than probably exists, and to press some very farfetched cases into his general symbolic scheme. Take, for example, what is perhaps the most celebrated panel in the whole of Paleolithic art, the painted ceiling of Altamira. This is dominated by fifteen large bison, with a few other creatures painted at the sides, including one horse. In Leroi-Gourhan's scheme the fifteen bison count as one symbol equivalent to the single horse, and so the ceiling forms a balanced composition, even though in reality the bison overwhelmingly outnumber the horse and capture the center of our attention. There is

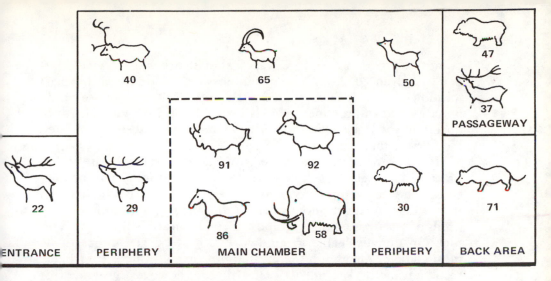

The "ideal" layout of a Paleolithic cave sanctuary, as determined by Leroi-Gourhan after a survey of sixty-two sites. The figures are the percentages of particular animals that he found in different parts of a cave. Many recent researchers disagree with this approach to the problem of finding order in cave art.

no real proof that the pairing of species is as regular and deliberate a feature of cave art as Leroi-Gourhan claims. If horse and bison were depicted more often by cave artists, then obviously the chances of the two species appearing side by side are higher. This element of chance was not properly considered by Leroi-Gourhan.

It has been said that Leroi-Gourhan's theory is a sexual interpretation of cave art, and he has been criticized for attributing "Freudian" preoccupations to the cave artists. This is incorrect, for the meaning of the twofold division of animals, signs, and humans is never stated to be a narrowly sexual one. Indeed, in a recent statement Leroi-Gourhan stated emphatically that he considered

> the dualism could be interpreted as an opposition of a sexual kind, but it could also have quite another meaning.[12]

Nevertheless his theory involves some very curious contradictions. The abstract signs are classed as male and female mainly on their resemblance to stylized male and female sexual organs. So all the "wide" shapes such as rectangles, triangles, ovals, and shieldlike forms are claimed to be symbolic vulvas, while "narrow" signs, strokes, dots, and barbed lines are considered to be phallic and male. Why such a bewildering variety of abstract signs should have been needed to express only "male" or "female" is not at all clear. Explicitly male and female human figures are, of course, classed in their appropriate categories, but the same does not apply to animals. Thus, if a bison is depicted with a phallus it is nevertheless considered to be part of the "female" group. Perhaps one complicated Paleolithic philosophy did embrace all these conflicting conditions. If so, it cannot be proved by the topographical evidence which Leroi-Gour-han has advanced. As long as it is impossible to be sure of the exact location of the original entrances used in Paleolithic times, or of the precise limits of prominent panels and compositions, an element of uncertainty will always intrude on any attempts to test so complicated a theory as this one.

However, to Leroi-Gourhan we owe an immense increase in our practical knowledge of the contents of the most important painted caves. This new information was conveyed not only by the detailed inventories and descriptions of caves, but by the publication of hundreds of unsurpassed photographs (taken by his colleague Jean Vertut). On this invaluable groundwork Leroi-Gourhan imposed the most complex fantasy on the meaning of cave art that is ever likely to be advanced. The impact of his theory, however misguided or far-fetched it may have been, is still felt by scholars and enthusiasts, for it opened up a new, more systematic era of cave-art study.

The most important milestone in this new phase of investigation was the appearance in 1967 of *Palaeolithic Cave Art,* written by two British prehistorians, Peter Ucko and Andrée Rosenfeld. Unlike Leroi-Gourhan's huge book, this was not a lavishly illustrated monumental work, nor was a single explanation advanced at the expense of other possibilities. Instead, the authors performed a vital service by critically reviewing the chaotic collection of fanciful theories and assertions which had accumulated up to that date. This involved a lengthy demolition of the elaborate structure which Leroi-Gourhan had sought to raise from his cave-art statistics. The authors systematically rejected a narrow interpretation of the art, whether based on symbolic pairing, or hunting magic or totemism, and instead insisted on the variety of contexts in which cave paintings are found and on the variety of possible motives in the artists' minds. The tone of their argument was very different from that of commentators half a century ago, many of whom had assumed that with the aid of information about the Aborigines or the Bushmen, we could confidently reconstruct the Paleolithic mind. The last sentence of Ucko and Rosenfeld's book makes this very clear, for they stress the possibility

> that some and perhaps many Palaeolithic representations were made for reasons which still totally escape the modern observer.[13]

On the other hand, they did not conclude that study of the art of recent "primitives" was entirely useless, as French investigators had done, including Leroi-Gourhan. In his pioneering enthusiasm for explanations based on inventories and statistics rather than those borrowed second-hand from the Aruntas or the Aleuts, Leroi-Gourhan had emphatically rejected the use of ethnography:

To take what is known about prehistory and cast about for parallels in the life of present-day peoples does not throw light on the behaviour of prehistoric man.[14]

Yet an over-reliance on the twentieth-century mind in interpreting the data from inventories and statistics led Leroi-Gourhan to an implausibly coherent view of cave art. At one point his inspiration, it seems, was drawn not from his inventory of cave art nor from the Australian Aborigines, but from nineteenth-century Europe:

> Palaeolithic Europe seems extraordinarily like present-day Europe or, to be more accurate, like nineteenth century Europe. A single religious system seems to underlie the works of art from Russia and the Ukraine to Spain.[15]

Ucko and Rosenfeld's work questioned such a unified vision of Ice Age art. Certainly, earlier writers had taken the records of anthropologists and used them to impose narrow explanations such as hunting magic, or absurdly detailed reconstructions (for example, one commentator explained the clay "sausages" found in the low chamber at Le Tuc as penis sheaths identical to ones used in Congolese initiation rites).

But a wide range of varied comparisons with recent hunter-artists could also be taken to counteract assumptions and ideas based on personal twentieth-century experience. As Ucko and Rosenfeld pointed out, it was illogical for Leroi-Gourhan to have done so much to broaden our knowledge of cave contents and yet to have insisted on narrowing our interpretation of this new data only to a present-day European viewpoint. A catholic view of the motives behind cave art means that we have much to learn from an equally wide consideration of primitive paintings, myths and rituals.

With Ucko and Rosenfeld's book, the grand theorizing of so many experts and amateurs over nearly a century was set in proper perspective. Single all-embracing explanations no longer provided a key to unlock the mystery of cave art. In recent years the most distinguished new research has concentrated not on the broad meaning of cave art, but with an ever-narrowing focus on the problems of recording even the most minute details of the caves and their decorations.

The outstanding figure in this new wave of research is Michel Lorblanchet, who has shown what dedication is required to extract the maximum information from even a modestly decorated cave. One of his reports involved 280 sessions of work underground, each lasting several hours, to record only a single wall of engravings about nine meters long, comprising fifty-nine separate designs in all. The plans of this one panel took a year to draw and involved the study of 1,500 photographic enlargements and numerous casts taken with silicone. The conscientious recording even of the faintest scratches can be vital, for these marks may sometimes be deteriorated designs which can be reconstructed only with the aid of different light sources and photographic blow-ups.

Lorblanchet's efforts illustrate the inadequacies of conventional cave inventories and the slippery basis of some past theories. For example, Leroi-Gourhan had called attention to the importance of the position and context of art works in caves, yet had given inadequate attention to how the forces of corrosion and concretion can obliterate entire walls, and so vitally affect the contents of the cave. Lorblanchet explains:

> It is the totality of the cave which must be studied, not simply the decoration. . . . The caves are not abstract entities. Each has a life of its own. The context in which the art works were placed must be "characterised." All this demands, as we have seen, a long time spent underground, a great expense of effort and the collaboration of specialists from different disciplines.[16]

Lorblanchet's work may not have thrown much new light on the motives of the cave artists, but he has shown how fascinating detail can emerge from unimpressive cave walls, and all of this is an aid to understanding. An immense amount of practical recording of this standard remains to be done before the paintings and carvings are lost forever.

Can worthwhile general conclusions be drawn from an art that is still so inadequately known? That there is order and purpose in cave art and that it is not just a random collection of Ice Age animal depictions seems certain. The fairly consistent selection of some species at the expense of others tells us this. Stevens' tentative study suggests that some of the order in cave art may correspond to regional tastes for particular themes or species. But what such selection really means in terms of creation myths or totemic histories or hunting magic, and how elaborate or widely understood such concepts were, is totally unknown. The abstract signs do hint at a complexity of ideas and hidden meanings. A small proportion of the animals were the object of "wounding," usually by unrealistic-looking arrows. Children as well as adults frequented the painted caves, but the concrete evidence for ceremonies is so inadequately preserved and recorded that it helps the inquiry little.

Beyond these generalizations, there is little to add, except to express the personal reactions to the painted caves of one observer, the present writer, which are no better than anyone else's. Each cave seems to have its own particular character, and one usually forms an impression of an individual hand or hands at work among the designs, which is perhaps not so surprising. Yet, contrary to this response, there runs a strong feeling of unity embracing not just the works of a single cave, but cave art in general. One cannot ignore this impression entirely by acknowledging the great variety of contexts in which the art is found or the variety of purposes for which Aborigines paint. Much of this is surely due to close similarities in conventions of style, which, as we have seen, were probably spread by the wide-ranging movements and contacts of Ice Age hunting bands. The great question remains as to how far this unity of style corresponded to a unity of meaning (if it is possible or worthwhile to distinguish the two). The question perhaps lies beyond the power of the prehistorian to answer, as Annette Laming-Emperaire concluded:

> We will never restore the living thought of the past. It is only a question of bringing the problem closer to us, without ever resolving it entirely.[17]

In Cartailhac's words of a century ago, the paintings and engravings "have a meaning which eludes us still."

11: *Venus and Beast*

The "Goddess" in the Jeweler's Shop

One of the oddest stories of archaeological discovery begins one day in July 1970, when a car full of French prehistorians was making its way through the narrow lanes which wind across the limestone plateau country south of the Dordogne. At some point the driver took a wrong turn, and the car arrived by mistake in the village of Monpazier, where the archaeologists decided to stop for refreshment. Glancing in the window of a jeweler's shop, one of the prehistorians noticed some flints on display. On entering the shop the group were told by the proprietor, a Monsieur Cérou, that he could show them other ancient tools and a "Venus" statuette which he had collected the previous April. The skepticism which greeted this remark—for prehistoric sculptures of any period are rare things indeed—swiftly changed to astonishment when Cérou produced a tiny female figurine which gave every appearance of being an authentic Paleolithic discovery.

A microscopic study of the surface of the statuette (sculpted from a conglomerate material) eventually confirmed that the carving was not likely to be recent. Meanwhile, an examination of the nearby field where Cérou claimed to have picked it up in a lump of earth showed that it had probably come from a disturbed living site of the late Perigordian period—perhaps 25,000 B.C. While washing the figurine, Cérou accidentally dropped it and it broke neatly into three pieces. After its repair a journalist tried to buy it (later, it again slipped through the fingers of a prehistorian and the feet were damaged). Such were the peculiar events before one of the most intriguing of prehistoric sculptures finally came to be studied by the experts.

This find from Monpazier is an example of a so-called "Venus" figurine. If the tiny chocolate-brown statuette were turned over carefully in the palm of a hand (it is only 5½ centimeters tall) you would be struck first, perhaps, by the protruding abdomen and the huge vulva sculpted carefully below it. Without arms or facial features except for the crude cup-shaped eyes, the whole figure seems concentrated on this swollen and distended region. The drooping breasts and upsurging buttocks are emphasized too, adding to the impression that the Monpazier statuette must have been a powerful image of pregnancy and sexuality. On the basis of this one figurine, we would probably not hesitate to conclude that Paleolithic people worshipped some sort of fertility goddess.

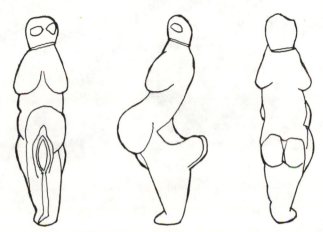

The "Venus" statuette from Monpazier, shown here at its actual size.

This notion of an all-powerful Stone Age Mother Goddess is a popular one and has influenced many interpretations of statuettes similar to the example from Monpazier. One reason why it arose was the attention paid to just one or two of the early discoveries (such as the figurine discovered at Willendorf, Austria, in 1908) which happened to be among the most grossly proportioned of all the female images. Their reproduction in countless books on prehistory has influenced the way we view Paleolithic statuary, or as a French expert put it recently,

> They have unceasingly solicited our eye and imprinted them-selves on our memory.[1]

The emphasis on the voluminous buttocks, long plump breasts, and swelling abdomen suggested to the early commentators that the significance of the statuettes was purely erotic (and so the inappropri-ate "Venus" name arose). When later discoveries complicated the picture and suggested that the range of proportions was considera-ble, it was still possible to argue that they were all connected in one way or another with an ill-defined concept of fertility. Here is how one of the foremost commentators on Paleolithic art, Paolo Graziosi, expressed the idea in 1960:

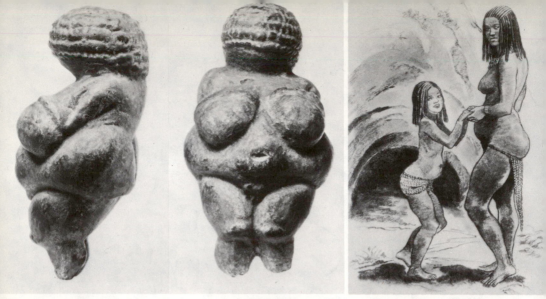

A cast of the Willendorf statuette and *(right)* a reconstruction of Paleolithic woman published in 1936 and based on an absurdly literal interpretation of the features displayed by the Willendorf and other stylized figurines.

> the exuberant fleshiness, full of life and feeling; the sense of female fulfilment radiating from the little masterpieces, unequivocally reveals that the end sought by the artist was an expression of fertile femininity, maternity in its fullest, most absolute sense.[2]

This temptation to explain the statuettes in terms of one fertility cult is still strong.

It is a striking fact that some features of the Monpazier statuette which appear so obviously sexual in our eyes occur very rarely on other Venuses. For example, the carefully worked vulva is present on only two or three other figurines like those from Willendorf and Grimaldi, and even on these few examples it is marked much more discreetly than in the Monpazier case. Other features, notably the distinctly carved feet and eyes, are almost unknown among Paleolithic statuettes. The true diversity present in the human representations is not, however, easy to grasp by personal impressions or by casually comparing one image with another. It has emerged only as the result of painstaking studies of the entire range of nearly 500 human figures known throughout Europe, including carvings and paintings as well as statuettes. These inventories of human figures, assembled quite recently by Ucko and Rosenfeld and notably by the Frenchman Léon Pales, have seriously challenged many past assumptions.

For example, the existence of a single ancient fertility cult seems a more unlikely proposition than ever. It would be easy enough to accept if the breasts and abdomen of female figures were invariably as exaggerated as they are in the case of Monpazier or Willendorf. However, the new studies show that representations with what might be termed realistically proportioned breasts and hips are actually more numerous than the familiar obese form. Among these slim figures are some striking works of art, such as the statuette of Laugerie Basse and the three torsos carved side by side in high relief on the walls of the rock shelter at Angles-sur-l'Anglin. The common problem uniting all Paleolithic human representations is not the question of why parts of the body were sometimes stressed, but why the legs and feet (and to a lesser extent arms and hands) were so consistently neglected, creating such a curiously unbalanced effect. This fact is all the more puzzling when we observe how often the hooves and paws of animals were depicted with care and accuracy.

The real surprise which emerges from the new work, however, is the high number of sexless figures—that is, figures which cannot be assigned to either the male or female sex based on the presence of obvious features like sexual organs or breasts. In western Europe many more sexless figures exist than either male or female forms. Perhaps these merely represent male or female designs that were hurriedly or incompetently executed. However, if Paleolithic artists really were obsessed by fertility or sex, we can imagine that their crudest sketches, like schoolboy graffiti, would consist simply of organs and breasts, and not what we actually encounter: a large number of anonymous and asexual human outlines. Clearly it becomes more and more difficult to maintain that the prehistoric

Strikingly realistic images of the female form do exist, like this ivory statuette from Laugerie Basse, here seen to best advantage in a greatly enlarged rear view. It is marked with a vulva on the front but no breasts.

artists were dominated by one ideal of the human form or by one cult image. As Peter Ucko has concluded:

> The figures represent a heterogeneous group of individuals rather than one single individual, human or deity.[3]

How do the little statuettes like those from Monpazier and Willendorf stand in the light of this general survey of human forms? Since Paleolithic imaginations were not overrun by pregnant Mother Goddesses, could these figures have some other, lesser meaning in terms of casual superstition or folk belief? In eastern Europe and Russia, where decorated caves or engraved stones are nearly absent, the human representations take the virtually exclusive form of statuettes, and at least seventy of these are known at present. The majority are female, although once again there is a healthy proportion (36 percent) of sexless figures.

Among the most remarkable discoveries were the partial or complete remains of thirty figurines found buried in the floors of six huts at Maltà, on the terraces of the River Belaya near Lake Baikal in eastern Siberia. This remote settlement developed its own idiosyncratic styles of flint-working and ornament, and the statuettes display some highly original features. The proportions of many of the Maltà figurines are realistic, and one face is depicted in a sufficiently individualized way as to suggest that it might even be a personal portrait. Several of the figures appear to be clothed in hoods, while other details of clothing are hinted at by simple arrangements of carved lines, dots and crescent shapes (Russian scholars claim to be able to recognize girdles, shoulder belts, and back packs from these markings). Several are perforated at the feet end, suggesting that they were meant to be hung upside down, perhaps as personal ornaments. Once again the variety of these eastern European figurines must be emphasized, although "fat ladies" do exist quite close to the familiar Willendorf pattern (for example, statuettes from Gagarino and Kostienki). Are all these figurines connected with one meaning and function, or with several?

The archaeologists can do little to help the inquiry, for the great majority of figurines in the west were dug up by amateurs without a proper record of their exact location or context. In eastern Europe the statuettes have usually turned up associated with the remains of ordinary huts—in several cases stored in special pits in the floor placed near the hearth. The most interesting find was at Gagarino in the Don valley during the late 1920s, where the partial or complete remains of seven statuettes were excavated, lying along the interior edges of a hut floor. This location suggests that they had once been attached to the wall.

Such clues have encouraged a number of Russian prehistorians to make direct comparisons with the figurines and myths of recent Siberian and Eskimo hunters, almost as if we could "read" the exact meaning of the Paleolithic sculptures from the beliefs of modern primitives. So a kind of composite picture of the ideology behind the figurines is built up, drawing on the myths of the Altai, the Aleuts, and many other tribes, and on folklore such as the Tungus practice of storing in each tent an image of the hearth spirit, who is portrayed as an old but vigorous and cunning woman. Taking a wide range of comparisons so seriously results in a very complex mosaic of Paleolithic beliefs; so we read that

> The image of the woman who was embodied in the female figurines reflected the important role of the Woman-Mother in the life of Upper Palaeolithic society. This role was interwoven with ideas of the woman as mistress of the home, hearth and domestic fire—the female ancestress with a special magical power which assured success in the basic source of life, the hunt.[4]

Some or all of these beliefs attested from modern sources may be true of the Paleolithic figurines. The danger in telescoping together so many superstitions drawn from ethnographic sources, as the Russian scholars have done, is that we may end up attaching undue social significance to the figurines, just as with the traditional Mother Goddess or fertility cult explanations. Could they just be amulets to bring good luck?

Beastmen or Sorcerers?

If explanations of every kind abound for the Venuses, it is much harder to account for the apparent reluctance of artists to depict human forms of every kind on cave walls and on bone objects, contrasted to the popularity of animal images. Only a handful of decorated caves contain human representations, comprising about 130 depictions in all, a small fraction among several thousand bison, horse, deer, and other creatures. Some have accounted for this rarity in terms of a religious taboo on depicting the human body, which would have been curious indeed alongside the practice of stenciling hand impressions known from several major caves.

Even more obscure is the fact that unrealistic bestialized forms are almost exactly as numerous in cave and mobile art as are realistic types. The majority of them are shown without any obvious sexual characteristics. The frequency of isolated animalized heads in particular has excited attention; there are about forty-two of these known from cave walls (as against only eight realistic ones), and these are

A curious humanlike figure engraved on a bone from Pin Hole Cave, Creswell Crags, Derbyshire, central England.

concentrated mainly in the two caves of Les Combarelles and Marsoulas. We cannot make assumptions about the sex of these often disturbing portraits, for we know of at least two mobile pieces on which breasts are shown on bodies joined to animal-like and apparently bearded faces.

There has been no lack of imagination in accounting for the curious postures and muzzle-like profiles which appear in some of these designs. For example, the psychologist Luquet suggested that they were simply the product of artists too accustomed to the creation of animal outlines to cope with the problem of representing the human form. Even more fanciful is the suggestion from a distinguished Spanish scholar, Magín Berenguer, that they represent racist caricatures by *Homo sapiens* of a genuinely bestial Neanderthal man, supposedly coexisting with the cave artists!

Many of the bestial figures, or "anthropomorphs," *do* seem to differ in quality and conception from the rest of Paleolithic art. It is often claimed that they are associated with one of the few widespread, recurring themes or scenes in cave art, specifically that of a wounded or threatened man. One or two cases of this theme are convincing. Several such wounded anthropomorphs pierced by darts can be seen in the caves of Cougnac and Pech-Merle, and all show curiously animal- or bird-like heads. However, despite a similar design on an engraved stone from the nearby site of Cabrerets, it is debatable whether the wounded-man theme extends further. There are at least half a dozen other scenes in caves and on bones in which humans appear to be confronted or pursued by animals; however, the details of each scene differ and none of the figures seem to be as unmistakably wounded as those from Cougnac, Pech-Merle and

One of several "wounded" anthropomorphic figures drawn on the cave walls at Cougnac, Dordogne. The figure is inside the large outline of *Megaloceros,* an extinct form of elk. At Pech-Merle, 40 kilometers south, the ceiling of a cramped nook bears this strange being *(below),* pierced by three darts and with an abstract sign emerging from the top of its birdlike head.

Gabrerets. Moreover, there are a far higher number of occasions on which anthropomorphic figures and faces are found in apparently peaceful coexistence.

One of the most remarkable of these designs was discovered in 1970 at the Magdalenian site of Torre near the northern Spanish coast, engraved with extraordinary precision around the surface of a narrow pelican bone a mere centimeter in diameter. The composition features six animals with elongated necks and expressive muzzles, and lurking below them is a carefully executed bearded bestial head joined to a stumpy body. What connection, if any, was intended between the beast and the realistic animals? We find it more attractive and easier to speculate about the dramatic confrontations such as the birdman in the shaft at Lascaux, but such scenes are exceptional. The peaceful and peculiar juxtapositions of anthropomorphs and animals are more numerous and more obscure.

The same problem hangs over the significance of the true beastmen, the composite figures which unmistakably feature parts of animals such as bison or deer horns. Among them is the famous Sorcerer of Les Trois Frères. Indeed, the attention paid to these undeniably strange beings seems quite out of proportion when only nine of them exist in the whole of Paleolithic art, and six of these are from only two caves: Les Trois Frères and Le Gabillou. Whether they were intended to represent hybrid deities or disguised shamans, the impulse behind their creation cannot have been widespread, and so one is again left with the question: what is the significance of such exceptional depictions and themes? Do they represent myths or only the whims of one artist?

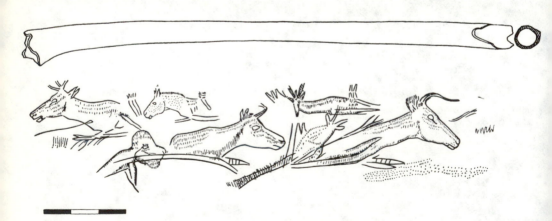

The finely engraved pelican bone excavated at the site of Torre, northern Spain, in 1970.

The Revelations of La Marche

Perhaps no site poses this problem more acutely than the rock shelter of La Marche, situated near the village of Lussac-les-Chateaux in western France, where prehistoric engravers seem to have delighted in breaking all the conventions of Paleolithic art. So extraordinary was the initial "decoding" of the La Marche stones, especially when touched up in a fanciful way by one of their discoverers, Stéphane Lwoff, that their original publication in the 1940s aroused widespread disbelief. After the presentation of certain of the human figures at the French Prehistoric Society in 1941, several members expressed

> their astonishment—indeed their stupefaction—and their doubts on the authenticity of the engravings.[5]

Today, thanks to the exhaustive investigations of Léon Pales, there can be no doubt that the La Marche engravings are genuine. There are too many of the limestone blocks—1,500 of them in all—for a modern forger to be responsible, and documentary evidence shows that the Abbé Breuil himself helped to supervise the excavation of many of them from an undoubted Magdalenian occupation layer under the rock overhang. Then there is the character of the carvings themselves, which is extremely peculiar: a mass of hairline scratches, a tangle of superimposed outlines engraved one on top of the other almost invisibly into the rock surface. It would have taken a forger of extreme persistence and eccentricity to have created such a jumble.

From the start it was clear that the subjects of the carvings were not the ordinary designs of Ice Age art. Breuil at once noted that there were several felines and bears, normally rare in cave art. He was particularly struck by

> four or five extraordinary male profiles of a very dignified type of human, with long aquiline or upturned noses: very fine indeed. Others are grotesque, including one reminiscent of a caricature of Hitler because of his snub nose and rebellious shock of hair over the forehead. . . .[6]

As the decipherment continued, mainly carried out by one of the original excavators, Stéphane Lwoff, it became clear that there were not four or five but dozens of strange human faces and figures intertwined with animal outlines or apparently meaningless random lines and scribbles.

The possible confusion in distinguishing the exact profile intended by the original carver was exploited to the full by Stéphane Lwoff. When he first wrote about the human figures in 1941, he claimed to have found on certain stones precise details of costume, jewelry and even complete figures which no one has ever been able to trace again from these same stones. Indeed, his report is a delightful piece of archaeological fantasy. He selected one head out of dozens, apparently misread the tangle of carved lines so as to produce an amusing caricature with a grotesquely protruding jaw, and dignified this one cartoonlike character with the title of "Lussac Man." Lwoff concluded that his improbable drawing could actually yield useful anatomical information proving that Ice Age man was a chaotic mixture of physical and racial traits, including an apelike jaw closely similar to Piltdown Man (at that time the Piltdown forgery, which incorporated an actual ape's jaw, had not been unmasked). While interpreting one of his heads in such a serious spirit, Lwoff was forced to admit that his delightful cartoon of a human head joined to the body of a penguin was "apparently, a caricatured man." Neither of these two figures has been found again.

The study of such an extensive collection of prehistoric "sketchbooks" obviously would require more than one inspired imagination and a casual examination of only a few stones. The task of deciphering all 1,500 of the blocks was a massive undertaking, requiring years of patient teamwork led by Dr. Pales, and involving ingenious techniques to sort out the true outlines of the figures. The blocks were examined with the aid of a magnifying glass, and then of enlargements of photographs taken with light sources aimed from a variety of angles. Finally, in some cases plasticine casts were made of the surfaces, since these provided not only accurate impressions for study but were also less of a strain on the eyes than the original stones. After all these efforts, the ambiguity of many of the figures remains, and in Dr. Pales' study several alternative versions of a single carving are occasionally offered.

The result of all this patient work is an intriguing Ice Age portrait gallery of fifty-seven isolated heads, with an additional fifty-one more or less complete human heads and bodies. These representations together constitute well over a quarter of all the human figures known in Paleolithic art, a fact which emphasizes how much the discovery of one extraordinary site like La Marche can transform simple conceptions about Paleolithic artists and their chosen themes. For instance, La Marche stands at least one traditional assumption on its head—namely, the proverbial reluctance to depict the human form, for at this site they are more numerous than any animal theme. Then the faces themselves contrast remarkably with the animalized profile so familiar from other sites. Far from appearing to represent

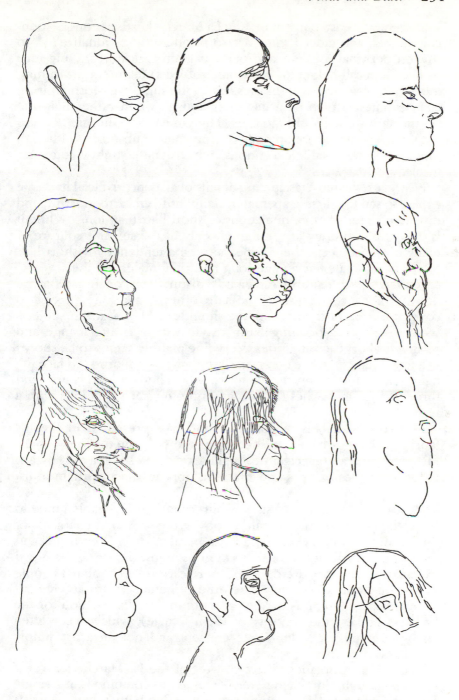

A small selection of heads from La Marche drawn for comparison at the
same scale and orientation.

some general beast-man notion, the La Marche heads are unquestionably human and indeed some are so peculiarly individualized as to suggest personal portraits and perhaps even caricatures. Only four or five faces (all belonging to female subjects) feature a protruding muzzlelike jaw, and even in these cases the high, well-formed foreheads suggest this to be a matter of artistic whim rather than the portrayal of some hybrid monster. The same artistic license perhaps accounts for the larger number of upturned snub noses (although there are some notable exceptions) which lend a slightly comic air to many of the faces.

However, the most fascinating details of all concern facial hair. We expect to see the hirsute apeman so familiar from cartoon strips, and instead encounter more depictions of short hair than long. While it is impossible to draw firm judgments about the sex and coiffure of many of the subjects, there appears to be a tendency for short hair to appear on male figures and long hair on females. Of the seventy-seven faces with features sufficiently distinct for conclusions to be drawn, only seventeen show the hair to have been long enough to cover the nape of the neck or the shoulders. Half a dozen subjects seem to sport a "pudding basin" style with a frontal fringe and relatively short back and sides. Very fine plaiting seems to be shown in a few instances, while on one strikingly well-proportioned face the hair looks as if it is drawn back into a bun. There are ten beards and three or four mustaches (in two cases the mustache coexists with a beard).

Most of the details of costumes and hats presented in Lwoff's fanciful drawings have vanished from the new, meticulous copies. However, at least one clear headband exists, and so do a dozen hats. A few examples of belts and bracelets appear, most of them indistinctly.

By itself an inventory of such details cannot be a reliable guide to Magdalenian fashions, nor can it convey the very odd character of the La Marche engravings. The large number of isolated heads and the comic spirit of a few of them might suggest the straightforward explanation that they are doodles, mere caricatures of actual people carved on stone to pass away the time. Whether or not an element of truth exists here, it is hard to account in a similar fashion for the headless bodies (the majority of them female), which occur often enough to suggest a deliberate concept and not an artist simply running out of carving space.

Looked at in isolation, the character of the La Marche figures is curious enough, but when several images are combined on a single stone the effect is sometimes extraordinary. No better example exists than the "Dancer" stone, which attracted attention from the moment of its discovery by Breuil in April 1940. Having washed and brushed the limestone block, the Abbé found a great male figure, waving his raised arms, appearing to invoke or exorcise spirits, represented by grotesque masks.[7]

Subsequent commentators embroidered on Breuil's description, one of them even suggesting that the dancer symbolized the fertilizing union of the earth and sky, despite the flaccidity of his penis.

The careful drawings and descriptions published by Pales have reduced the number of figures or spirits on the stone to only two, and make the meaning of the design all the more obscure. The central dancer may not be dancing at all, for (like so many other Paleolithic human figures) the treatment of the legs is skimpy and not very clear. On the other hand, the great care with which the extended fingers and the facial features have been carved is highly unusual. The head, with its open mouth and bristle of hair at the back, is oddly expressive, but whether of anger, fear or exaltation it is impossible to say. The dancer is either lifting some object—nothing more than a confused tangle of lines can be distinguished—or else he is simply menacing the second figure leaning forward (or falling?) below his elbow. This second character, with arms extended and (less clearly) with knees bent double, has one of the more cartoonlike of the La Marche faces, and sports a goatee beard. There is nothing quite like this strange scene in the whole of prehistoric art, although another stone from La Marche bears some less distinct figures with arms raised high over their heads, obscured by a dense jungle of incomprehensible scribbled lines. Whether these gestures are really meant to represent dances, incantations, acts of violence or ritual, they seem to be another instance of the unusual themes selected by the La Marche artists.

Until the stones were published, it was possible to argue that no convincing depiction of human or animal copulation existed in Ice Age art. Once again, although the evidence is ambiguous, the La Marche engravers seem to have set a precedent. In over half a dozen cases, indistinct outlines can be interpreted either as single individuals of monstrously obese proportions or else as two intertwined people viewed from the side. While the theoretical positioning of the genital areas makes a sexual interpretation possible, the carvings are anything but explicit. A rapid sketch of one of these stones by Breuil himself survives, with the dilemma posed by it written out succinctly in longhand below: "Is this a fat woman seen from the front? or a coition?"[8] Inclining toward the latter view, he nicknamed this group of representations "marriage certificates."

The carving on at least one stone is clear enough to suggest that Breuil's preference may have been right. This stone shows a figure with arms upraised, appearing to clench her fists together in front of her face. The body is generously proportioned, especially the arms which seem to end in bracelets of some sort, while the head may be

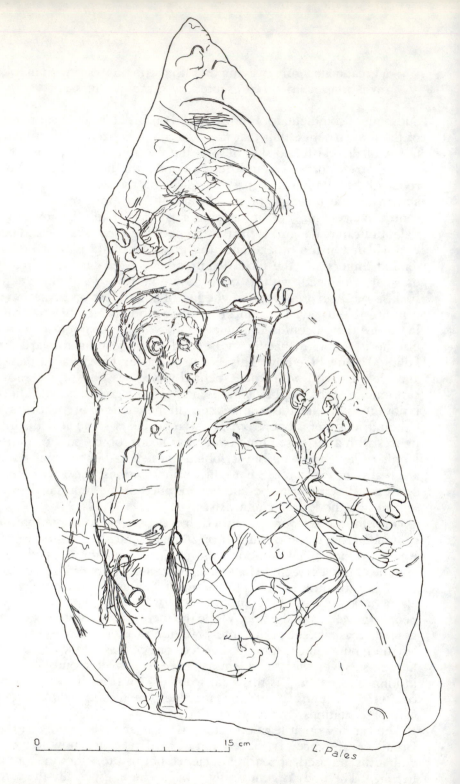

0   15 cm

L. Pales

Dr. Pales' copy of the strange scene on the "Dancer" stone from La Marche.

A close-up of the head of the "Dancer" from La Marche.

covered by a bonnet and the waist with a belt. Taking this carving at face value, we would probably not hesitate to view this individual (who has a female air about her although no breasts are shown) in a kneeling attitude of prayer or adoration. This would ignore, however, the faint carving of a second, much leaner figure facing in the opposite direction, again without any obvious sexual characteristics. The second individual also has an arm upraised, and is leaning forward with one leg flexed so that both bodies touch and the heads are superimposed. What can this addition of a second figure mean? Should we in fact turn the carving around 90 degrees to view the woman on her back in a position of intercourse? And could the absence of sexual characteristics be due simply to the artist's depiction in profile of an act during which they would, in fact, be invisible?

The temptation to find sexual meanings where none may exist is always strong, and in any case the peculiar "marriage certificates" clearly do not conform to any simple idea of a fertility cult or sexual magic. Instead, their ambiguous quality hints at something odder, perhaps even a reticence toward the explicitly erotic that we might assume from the absence of animal copulation and sexual organs from so many Paleolithic designs. In this case, as in so many others,

One of the strange "marriage certificates" from La Marche, apparently showing two human figures superimposed on one another. At right, the two figures are drawn separately for comparison.

La Marche has qualified and sharpened our awareness of the strange conventions of Ice Age artists, and thrown new light on some celebrated works.

Consider, for example, perhaps the most haunting of all the prehistoric human compositions, the remarkable carving on a small piece of bone (only 10 centimeters long and 6½ centimeters wide) known as *La Femme au Renne.* Along with other carvings from the same site, this was the earliest human figure ever discovered, dug up from the massive deposits at the Laugerie Basse shelter in the Dordogne between 1867 and 1868. Not long afterward, the carving was acquired by Edouard Piette, who noted the care with which a recumbent figure had been carved in relief under the rear quarters of a reindeer, without obscuring or crossing the animal's hooves. This suggested to Piette, and to most subsequent commentators, that the pairing of the "woman" and the reindeer was a deliberate composition, and that the two subjects were intended to be viewed as one scene. Piette was quick to interpret the scene as an act of bestiality, and then extended his interpretation to arrive at one of his more eccentric arguments in favor of the domestication of the reindeer: the recumbent woman was a symbol of the techniques of sexual arousal devised by Paleolithic hunters to lure and control the reindeer herds.

A careful scrutiny of the piece suggests that the composition is altogether more obscure in its meaning than any facile guesswork of this sort. One may begin by noting the detail with which the reindeer hooves have been executed and the "classic" neglect of the legs and feet of the recumbent figure. Then there are peculiar markings on the swollen tummy of the apparently naked figure, which make it look hairy but might simply be decorative. Just visible at the broken edge of the bone are three or four beads of a necklace, while the upraised arm seems to be encircled by bracelets. The most recent tracing of the piece, published by Pales in 1976, shows a discreetly figured vulva but no breasts, so whether we should regard the depiction strictly as a "Femme" is an open question. Finally, the new tracing shows that after the laborious relief carving of the reindeer

"Is this a fat woman seen from the front? or a coition?" wrote the Abbé Breuil, perplexed by this jumble of engraved lines on a slab from La Marche.

and the figure was completed, abstract rectang 'ar shapes were scratched lightly on the left-hand side of the composition, and even more peculiar sketchy outlines were added over the "hairy" tummy. Is this very inconspicuous design meant to indicate a partner engaged in intercourse with the "Femme" below?

Without insisting on it, Pales points out the parallel with the "marriage certificate" already discussed. Like the supposed "kneeling woman" of La Marche, the "Femme" of Laugerie Basse has no conspicuous sexual parts, its bracelet-covered arm is upraised and it seems to be involved in some extremely shadowy and indistinct contact with another human figure. So, discarding the simple notion of bestiality, one arrives at a more intriguing possibility: The image seems to link a reindeer with a scene of *human* intercourse, carved in two different techniques and perhaps at two different times. If the carving represents an act of sexual magic or illustrates a myth of reproduction, it is a more complicated one than Piette and many others ever supposed.

After exploring the remarkable diversity present among nearly 500 Paleolithic human figures in vain for some consistent signs of a fertility cult, we are left with a mere handful of images, these ambiguous scraps of evidence from La Marche and Laugerie Basse, which point intriguingly toward sexual beliefs and practices.

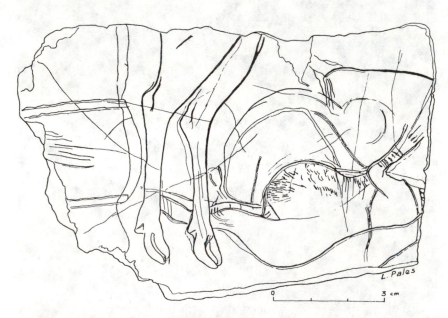

Dr. Pales' tracing of "La Femme au Renne," the engraved bone found at Laugerie Basse in the late 1860's.

12: The Mysterious Engravings

The Strange Sanctuaries

The people of the La Marche rock shelter could scarcely have chosen a more favorable spot for their activities. The entrance to the overhang was just high enough to stand up in, so that it could have been closed off easily in rough weather with branches or hides. The south-facing porch looked out over a slope running down to a stream a stone's throw away which had worn down a gentle valley through the limestone plateau. Down by the stream the ground was often damp but rarely swampy, encouraging the growth of alder. Closer to the shelter stood groves of lime trees, while in the stonier areas of the valley there grew a mixture of oak, box, and ash, fringed with bracken. In spring and summer the grassy slopes outside the shelter were fragrant with wild pansies, bellflowers, garlic, and oregano, and there were hazelnuts and red currants to be picked. This sheltered valley, rich in grasses but with the cover of trees never far away, must have attracted large numbers of animals down from the surrounding prairie.

Such a setting (reconstructed from the evidence of pollen blown or carried into the rock shelter 14,000 years ago) would surely have provided an ideal living base for a group of Ice Age hunters: abundant water, shelter, cover for hunting game, and a variety of edible plant foods. Certainly the size of the oval living space protected by the overhang (a mere 20 square meters) would not have permitted occupation by many people, perhaps no more than a small family. There is indeed evidence of such an occupation in the form of tools worked from flint, bone and antler, animal remains (especially horse teeth), and two hearths. There is at least one clear indication that the settlement at La Marche was of an intermittent or seasonal kind. Throughout the period of human activity at the site—probably in the late 12,000s—lemmings, dormice, field mice and voles burrowed into the low crevices at the rear of the cave, suggesting that La Marche was occupied periodically or perhaps only during the daytime.

The interior walls of the cave were much too rough to have provided a surface for decoration, so the engravers took portable slabs from a layer of better quality limestone higher up the valley, and during the course of an undefined period left more than 1,500 decorated stones on the floor of the shelter. Unfortunately the original excavators' records are far from complete, but it seems as if the soil was strewn with tools and engraved plaques all jumbled together throughout the space enclosed by the cave. Some plaques showed evidence of damage by fire and of being stood upright; others had been broken and the pieces dispersed, though whether deliberately or not can only be guessed. The general impression seems to have been one of disorder and perhaps negligence on the part of the artists. One has the curious picture of engraved stones being trampled underfoot in the half-light of the confined overhang, especially if the occupants had tried to pick their way round to the back of the shelter with its low, sloping roof. One could imagine the plaques propped upright, perhaps even on wooden stands, except that attempts to exhibit and store certain stones in exactly this way in a Paris museum have proved impracticable. If La Marche was a sanctuary, then it cannot have been an impressive one and seems to have been treated rather carelessly; but if it was not a sanctuary, how are we to explain the astonishing character of the carvings left behind?

The delicate engravings are not only of unusual subjects but some are of unusually high quality, like the ferocious lion heads discussed in Chapter 10. Yet many of the finest representations are badly obscured by other outlines superimposed in an apparently haphazard fashion, or else by a mass of meaningless scribbles and "parasitic" lines. Certain of the plaques bear traces of red ocher, and it is possible that a colored wash was applied to the stone on each occasion to cover up earlier images, although this would not account for the depth and detail of many engravings. Nor would it explain the frequent choice of the roughest and most irregular side of the stone as a carving surface, a fact which perplexed Pales and his collaborators. "Everything looks to us," he wrote, "as if the carver had deliberately made things difficult for himself."[1] One explanation, which would make the engravers appear a little more rational in our eyes, is that some stones could have split apart between the time of carving on the smooth face and subsequent engraving on the rough side. This still does not really help to explain the choice of such unsuitable surfaces, nor the obscuring of so many designs by the "parasitic" lines.

A small number of other "plaquette sanctuaries" exist, which suggests that the practices of the La Marche artists were unusual but not unique. The first of these sanctuaries was brought to light in 1909 when two prehistorians, Louis Capitan and Jean Bouyssonie, began investigating a deposit which lay partially under houses and lanes close by the junction of rivers Vézère and Dordogne. At this spot, the village of Limeuil, it was apparent that Magdalenian occupation layers had been washed out of a shelter in the cliffs high above, and debris had tumbled down onto the river banks below. Among the disturbed material recovered were 137 limestone blocks with clearly identifiable animal engravings, and many more with abstract, incomprehensible patterns. Some of the stones showed marks of burning.

Once again it seems that the Limeuil artists were nonconformists, concentrating their skills on subjects which are rare in cave art, including more than fifty lively depictions of reindeer in naturalistic poses such as grazing and galloping. The engravers at Limeuil did not depict humans nor any bison, while the rarity of weapons or wounds in the animals, plus their vitality, suggested to the excavators that art was uppermost in prehistoric minds. A slightly later commentator thought that the naturalism of the Limeuil animals altogether ruled out superstitious motives:

> Imagine the indignation of the austere medicine-man, who caught a youthful brave taking in this easy-going fashion a task that should involve a gloomy ritual and an atmosphere of religious terror.[2]

The excavators further noted a curious fact. Heads and feet were sometimes carved repeatedly in slightly different positions, as if the artists were seeking a perfect outline or as if novices were trying out details of the design for themselves on the same stone. So was born

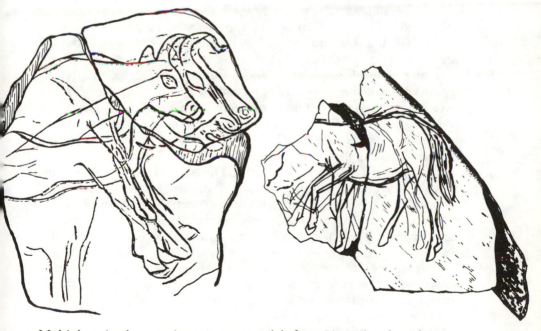

Multiple animal engravings on a stone slab from Limeuil and *(right)* on a bone from Laugerie, both in the Dordogne region. Not at the same scale.

the theory that Limeuil represented a Magdalenian "Royal Academy," a Paleolithic art school, where pupils were taught by masters before being allowed to decorate the supposedly sacred cave walls. For Capitan and Bouyssonie, the stones represented nothing less than "album leaves" of sketchbooks, designed to perfect the careful eye for anatomical detail so evident from the best of the engravings.

The theory presented a comfortable explanation for the excellence of so much Paleolithic art, and rapidly grew in popularity. The year after the publication of Limeuil in 1924, one writer described prehistoric artists in terms of a Paris studio; the gifted

> formed themselves in select groups, just as today's famous artists form their circles, and they were the masters; their pupils, the product of competition, gathered round them; then with the death of the most talented . . . the earliest art school . . . which had begun so perfectly in the hands of several or perhaps only one genius, dissipated itself little by little and sank into the obscurity from which it has been rescued by modern science.[3]

One characteristic that we would expect of an art school would be the recognition of the hand of its master or pupils in places other than the studio itself. However, it is the distinctiveness of the Limeuil, and especially the La Marche, engravings that impresses us, not the very general resemblances which they show to other examples of cave or mobile art. The unique preoccupations of the La Marche "sketchbooks" (if that is what the engraved stones are) surely rule out the idea of a training school or studio in the modern sense.

In any case the next "art school" site to be discovered after Limeuil knocked the entire theory on the head. From 1929–31, the cave of Parpallo in Valencia, situated close to the Mediterranean coastline of eastern Spain and hundreds of kilometers from the major painted caves of France and Cantabria, yielded up no less than 4,983 engraved blocks distributed throughout several thousands of years of occupation by Ice Age hunters. The special conditions of the Parpallo site preserved traces of painted designs on the stones as well as engravings. These paintings tended to occur in the earlier period of the cave's use—they appear on quite large blocks found near the hearths around which the Solutrean hunters must have gathered for warmth. Later, however, engraving was more common and the smaller stones selected for decoration were treated much more casually by the Magdalenians, so that in some cases the plaques actually served as hearthstones or were left among piles of rubbish.

Throughout this long period of activity by the Parpallo artists, the familiar profiles or fragments of animals appear, but abstract patterns and indecipherable scribbles are equally numerous. The most important aspect of these designs, however, is their mediocrity. Not a single one of nearly 6,000 decorated faces would be impressive enough for a coffee-table book on Paleolithic art; almost without exception the animals are stiff and sketchy. Clearly an "art school" theory here would be most unconvincing, not only because of the lack of a master's hand among the designs but also because of the extreme scarcity of decorated objects within hundreds of kilometers of Parpallo itself.

The most recently discovered of these plaquette sites is equally far removed from the main centers of cave painting. The site of Gönnersdorf in western Germany came to light in 1968 when house builders were digging a foundation trench on a gentle slope overlooking the Rhine some fifteen kilometers northwest of Koblenz. Here, protected by the debris of a volcanic eruption which had devastated the region about a century afterward, the living floors of two large stone-walled huts were uncovered, dating to about 10,400

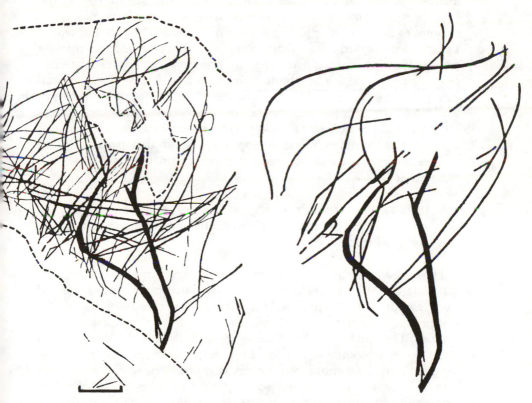

Engravings on a slate fragment from Gönnersdorf, western Germany. In the simplified drawing of the same slate *(right)* seven stylized overlapping figures can be seen, including one (upside down) figure with a pointed arm-stump and breast below it.

B.C. A pavement ran between the two houses, and outside them were two small rings of stones which had probably supported tent structures. Inside the largest house (some six meters across) there were two hearths and the floor had been colored red with powdered ocher. There were also storage pits containing personal ornaments such as forty wooden beads, pierced fox teeth, and seashells from the Mediterranean, while another contained pieces of ivory, bone and antler carved in strange abstract shapes.

The greatest surprise of Gönnersdorf came, however, when the pieces of slate paving the entire living surface were examined. Some 500 of these were covered with engravings, many of which represented mammoths and horses. Over 200, however, featured simple schematic female or sexless outlines, sometimes reduced only to legs and buttocks. This is the shape which seems to be repeated in the little sculptures found inside the large hut and elsewhere on the site. All these representations are shown without heads or feet, while the arms, when they are indicated at all, seem to be raised in front of the body. Seven of the figures are carved with stripes, which may be an attempt to depict clothing. There are many sexless figures, but where breasts are shown, their shape suggests maturity and an upright posture. In eight cases a pair of figures confront each other, but since only two of these feature bodies overlapping in any way, a sexual interpretation seems highly improbable. More frequently the figures are shown in procession, one behind the other, with the images sometimes separated so as to suggest a frieze or composition, and sometimes accumulated one on top of the next almost to the point of invisibility. Of the latter cases the excavator, Gerhard Bosinski, wrote that

> it looks like a half-dreaming person engraved these figures in rapid succession without concern for problems of space and composition.[4]

Half-dreaming or not, the engravers were content to let their slates be dropped underfoot, broken in many cases, and dispersed over a wide area. Sometimes the plaques were reused, so that an image was carved across a break which had interrupted an earlier outline. The quality of the slate itself—exactly like the slates which were once in common use for arithmetic in shops and schools—suggests that the complicated images *might* have been carved not "in a rapid succession," but over some period of time. Each fresh line carved into the stone creates a white dust, making it stand out sharply against previous outlines, which may be nearly invisible if the slate has been washed beforehand. So, were some of the humans carved at intervals, perhaps to fulfill some rite or superstition?

One further site, the great painted cave of Labastide in the Pyrenean foothills, may one day throw further light on these carving practices, but the results of digging just after the last war have not yet been adequately published. As a result we do not know exactly what was carved on the surfaces of nearly 1,000 engraved plaques excavated in a spacious gallery nearly 200 meters from the cave entrance. The walls of this gallery are ornamented with paintings and engravings, including a magnificent life-sized polychrome horse. The majority of the carved stones were found in fragments, although some intact ones surrounded a hearth with occupation material unearthed in the very farthest reaches of the cave, nearly 400 meters from the entrance. Perhaps if the full details of the Labastide discoveries were known, they would add much to our picture of the ceremonies or rites in which the engraved stones may have been involved.

Five sites—Labastide, Gönnersdorf, Parpallo, Limeuil and La Marche—were centers of an intense and peculiar artistic activity, concentrated on small pieces of stone. However, the artists were widely separated in space and in time, while their preferred themes at each of the five sites were entirely different. It may be entirely misleading to look for a common belief or practice—sacred or profane—which

Stone slab from the late Magdalenian site of Puy de Lacan showing a bird, the front quarters of one bison and the rear of another.

would link them all. Nevertheless, at each site certain curious conventions prevailed—notably, the practice of engraving fragments of animals as well as whole figures, and of superimposing outlines, sometimes to an almost indecipherable extent. To make any sense of this overengraving, we must suppose that successive images were added to the surface over an interval of time and not all at once. Finally, the decorated stones often seem to have been treated in an apparently casual way, abandoned in the mud and debris of each site as if they were worthless, or at least important only at the time of carving.

The Moon and the Bones

These activities were concentrated at a handful of sites like La Marche and Limeuil—suggesting, it may be, the existence not of "art schools" but of "sanctuaries"—but were certainly not exclusively confined to them. The superimposing of one painting or engraving on another is a common feature of cave art, and it is only on very rare occasions that the positioning of later works seems to bear any reference to what went on before. (It is at Lascaux that apparently self-conscious composing or pairing of animals seems most evident.) Considering the poor illumination afforded by their grease lamps, it is easy to imagine cave artists concentrating solely on the image in the narrow circle of light immediately before them and simply disregarding previous works. However, there are a number of caves in which certain panels are supercharged with engravings to the point where individual designs are lost in a chaos of hooves and manes and horns, despite the presence of suitable virgin surfaces close at hand. Two particularly notable examples are engravings in the Sanctuary of Les Trois Frères deciphered with such difficulty by Breuil, and stretches of the low ceiling in Gargas in the Pyrenees, where identifiable animal outlines are outnumbered by incoherent sketches and fragments of beasts such as horns and manes.

At the lowest level of interpretation, we can regard these surfaces as no more than "blackboard graffiti," one carving attracting another in a completely casual manner. Another view is that such panels represent clear evidence for the "magic" significance of cave engravings: the appearance was trivial compared to the act of renewal of sacred surfaces and images over a long period of time. Whatever level of explanation is preferred, the practice of accumulating chaotic engraved images known from a handful of caves seems to resemble the activities at La Marche, where the rough walls of the shelter itself would not have permitted carving.

In trying to fathom the mystery of these multiple engravings, the stone blocks and the cave walls present limitations. It is usually

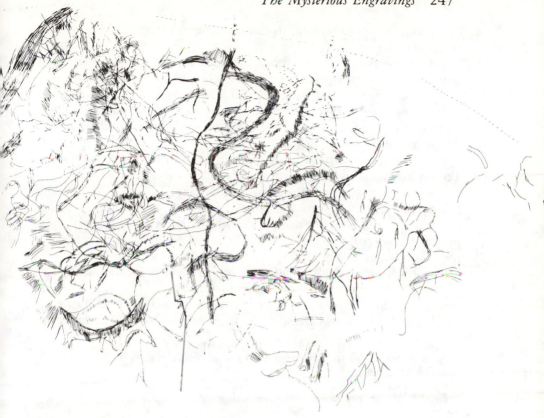

Detail of the labyrinth of engravings in the Camarin chamber at Gargas in the Pyrenees.

difficult or impossible to know which image came first or whether some period of time was really involved. With mobile art, and especially the carved bone and antler pieces, these problems can be attacked. By studying the miniature art in more detail than anyone had attempted before, one American researcher, Alexander Marshack, has succeeded in contributing many intriguing new ideas during the past decade, and has made some remarkable claims.

One of the most important of Marshack's propositions is that decorated bone tools and objects from all periods of the Upper Paleolithic bear traces not merely of the successive additions of one image on top of another, but of the actual *recording* of intervals of time. This claim is such a startling one that it requires a careful scrutiny. If Marshack is right, then Ice Age hunters were living in a far more ordered world than has been suspected up till now—a world governed by carefully fixed seasonal rites and observances relating to the simplest of calendars, the waxing and waning of the moon. At once the superimposed animals take on a new possible meaning: could they have been carved at significant moments, days of seasonal rites or sacrifice, to commemorate points in this elementary calendar?

To our eyes the carved bones would not look like recording devices, for there are no numerals or obvious symbols of the sun and moon. But there are rows of insignificant-looking notches carved into the edges of objects such as the mysterious batons, and these markings appear from the earliest centuries of the Upper Paleolithic down to the latest phases of the Magdalenian (when they often appear associated with carvings of animals and other abstract signs). To the antiquarians of a century ago these linear engravings were *marques de chasse* ("hunting marks"), a tallying system of some kind, raising the slightly absurd notion of a hunter scoring one point for every mammoth brought down. To others the notches were simply purposeless decorations, ornamental fringes or corrugations to assist the grip of a hunter on his tool.

Marshack encountered these markings not as a professional prehistorian, but as a journalist engaged in writing a book about space travel and the history of science. During research for his book in 1963, his attention was attracted by a photograph of distinct groups

Engraved bone objects, from the Abri Lartet near Les Eyzies *(third from left)* and from the Abri Blanchard, Sergeac, Dordogne. According to Marshack, the sequence of marks in the center of the second bone involved twenty-four changes of tool points or style, suggesting the object's use as a long-term "notebook."

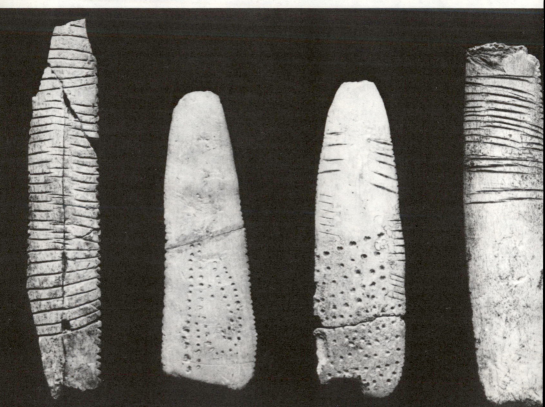

of linear markings on a prehistoric bone, and the theory that these could represent recorded events in time, or "notations," was born. Marshack described how his initial excitement "dizzied and burst in me like a fever."⁵ Weeks of research followed, until he decided to leave America to study the notches on Paleolithic bones at first hand in the museums of Europe, taking with him a camera and a Japanese toy binocular microscope. It soon became clear that microscopic study of the carved lines—especially of their cross-section and depth —revealed more information than could ever be gained with the most painstaking scrutiny by eye alone. The examination of the carved notches showed that the flint engraving tools had been changed many times in the course of any one sequence, suggesting that the markings had been created not with one tool on one occasion but at numerous separate moments. This simple observation, however we may interpret its significance, is a fundamental discovery and one that raises many questions.

For instance, we may wonder if the changing of a tool point could have any possible significance for a Paleolithic engraver. Was he able to recognize a change in the markings? Clearly, if Marshack was only able to discover some of these minute variations with the aid of a microscope, it seems fantastic that Paleolithic man should have used them to discriminate points in a tally or calendar.

Marshack has answered this in two ways. Firstly, that if one is familiar with a convention of marking, then the use of tiny graduations presents no difficulties, and he cites the minute marks on a watch or the letters in a finely printed text as examples. Secondly, the recognition of a change as each set of markings was made might not have been important to the original carver. Perhaps it was just the sum total that concerned the engraver. Furthermore, he may not necessarily have needed (or been able) to think of that total in terms of numbers or of an explicit counting system. The over-all sum or appearance of the tally might have been recognized without the use of numbers, just as an abacus can be used to encourage and express a child's basic awareness of quantities.

These replies might explain why Paleolithic carvers felt obliged to record their successive notations at such a tiny scale; next, we need to know why the notations were being made in the first place. If the structure of the markings was entirely random, then we could envisage the notations either as some very eccentric practice of decoration, using many different tools, or else as a tally of some sort, corresponding to the old "hunting marks" theory. In either case we could not hope to understand their meaning further than these initial guesses. However, if the number of marks or notches conforms to any pattern or regular rhythm, then it is at once reasonable to imagine that they form a record of some recurring event or interval of time.

This is the substance of Marshack's claim, and he extends it further to suggest that the repeating numbers which he has noted among the carved bones correspond to the simplest of all celestial regularities, the waxing and waning of the moon. Without recourse to complicated numbers, the carvers could have observed and recorded intervals between distinctive phases such as full and new moon, the quarters or crescents, with the object of keeping track of the lunar months.

Such a system appears to underlie all the earliest recorded calendars from the civilization of China to those of the Near East, while anthropologists have reported ways of recognizing and naming months from a great diversity of recent hunting communities. For such people an accurate time-reckoning system, involving a precise knowledge of the length of the year, is irrelevant. It is sufficient to know what month one is in and what the next will bring in the way of natural phenomena, such as the weather, the appearance of vegetation or the migration of animals. Many tribesmen—the Greenland Eskimos and the Indians of British Columbia among them—did not bother to fill up the year completely with recognized months; other peoples, such as the Dyaks of Borneo and the Algonquin Indians, kept a continuous lunar calendar, but had no idea how many "moons" constituted a year. It was the short-term reckoning that mattered, and the signs which gave titles to each month. The Koryaks of Siberia, to take just one example, recognized individual months associated with the cold winds, the growing of spinal sinews on reindeer, the calving of reindeer, the reddening of leaves, the rutting of mountain sheep, and so on. Events such as salmon runs, the opening of flowers, the thawing of rivers and lakes, the shedding of horns, cuckoo calls, and the movements of crows and geese, were adopted for rough lunar time-keeping among a vast range of recorded hunting peoples from the Arctic circle to the shores of the Atlantic. Marshack has supported his ideas by referring to recent calendars of the Nicobar islands, kept on carved sticks with an appearance almost identical to the Paleolithic engravings. There is nothing at all unreasonable in the suggestion that Upper Paleolithic hunters regulated their activities by recording the phases of the moon. The question is: Did they do it?

To *prove* that the recurring sets of numbers on the Paleolithic bones correspond to lunar phases is not so easy as it might seem. The moon is a constant and regular timekeeper, but its monthly cycle does not, of course, correspond to an exact number of days (instead, it is equivalent to roughly 29½ days). Without the benefit of this modern calendrical knowledge, an ordinary observer would find himself counting a month as anything between 28 and 31 days. Even greater variations would be introduced, depending on whether the observer began or ended his observations with the 2- to 3-day period of invisibility at the new moon, or started the reckoning on the exact day or the day after the appearance of the first crescent. Furthermore, uncertainties of observation or periods of bad weather would mean further variations in the lunar tally. If we accept, as Marshack does, that the observers were not only recording whole months but also halves and quarters (so accounting for the frequency of numbers such as 13, 5 and 8 among the notations), then the potential of proving *any* string of numbers to be lunar is considerable. Only in a long sequence of marks covering several months could we expect some of these factors to balance out and a possible lunar rhythm to be detectable in the markings. This is exactly what Marshack claims to have shown with a few remarkably long notated sequences, such as the eagle bones from Le Placard.

With all these problems in mind, it is not surprising that some researchers are wholly skeptical of attempts to prove that the lunar explanation is correct. Marshack has never published analyses of a large number of objects which would allow anyone to repeat the statistical tests which he claims back up his theory. Instead, his most detailed study is of only six objects with notches that *do* fit the lunar phases to varying degrees, which is not to say that another explanation might not fit them better. No one would deny the attractiveness of the theory of a Paleolithic lunar calendar, its convenience as a rational explanation of the "hunting marks," and its widespread use by peoples at a comparable level of technology. Common sense seems to tell us that Marshack is right, but as yet the proofs are lacking.

A Horse with Three Ears . . .

In any case, Marshack's work is not limited only to the controversial decoding of these puzzling abstract marks. He was responsible for a remarkable popular book, *The Roots of Civilization* (published in 1972), written "chapter by chapter, on the run, on planes and in hotel rooms, as the research enlarged and spread through the museums of Europe."[6] The text of the book contains many farfetched claims and wild interpretations, and subsequently the author himself has admitted its shortcomings. But few have criticized Marshack's presentation of close-up photographs of the engraved bones and stones—photographs of striking quality which introduced the intriguing scenes of mobile art to a very wide audience. Indeed Marshack's greatest achievement, perhaps, has been to make us look at a few of these remarkable decorated objects with new eyes, and to make us aware of their unexpected complexity.

One such object was the baton dug up at the site of Montgaudier in the Charente in 1885, which bears the surprising design of a male and a female seal—surprising because Montgaudier must have been even farther from the coast during Magdalenian times than its present location, one hundred sixty kilometers from the Atlantic. This is one more piece of evidence that journeys to the coast formed a regular part of the movements of hunters in certain regions. However, Marshack prefers the possibility that seals could have been seen far inland if, as happens occasionally today, they pursued a salmon run a long way upriver during the period of the spring spawning. Marshack lends weight to this suggestion by noting the tiny hook protruding from the lower jaw of a fish carved beside the seals, a characteristic feature of male salmon during the spawning season.

Further evidence that the Montgaudier composition might represent a symbol of spring is provided by outlines engraved on such a tiny scale that the microscope is essential to their decipherment. These little schematic designs—one looks like a budding flower, another is an ibex head crossed out with an X—look as if they were hastily tucked in among the larger realistic carvings as an afterthought. Marshack's idea is that the Montgaudier baton had some long-term use, and that in effect the carving of each image over some period of time added bit by bit to its over-all significance as a kind of pictorial "diary" of the coming of spring. The idea seems plausible enough, and suggests why some unusual depictions—such as the seals, the salmon, and the budding flower—may be associated together on one object. But the specific seasonal relevance of the other motifs—three odd insectlike creatures (Marshack suggests they are slugs), a pair of grass snakes, the ibex head, and some abstracted stems of vegetation—is less obvious.

Such complicated scenes are rare in mobile art (only about half a dozen truly comparable decorated objects exist for the whole of northern Spain, for example). Whenever they have been excavated

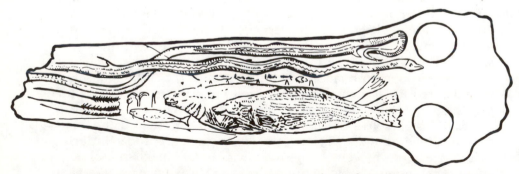

The design engraved on the rounded surface of the baton from Montgaudier in western France, "rolled out" in one drawing for clarity.

A plantlike sign painted in red at the cave of Marsoulas in the central Pyrenees.

with care, the complex scenes date from the very last stages of the Upper Paleolithic. To summarize it crudely, Marshack's work suggests that by this late period the Ice Age hunters were waiting and watching for the appearance of specific natural phenomena, sometimes mythologized or re-enacted in ceremonies, and sometimes measured in time quite accurately with the help of the lunar notations. In other words, the outlook of the hunters was structured by stories linked with natural events and the behavior of animals in time. Although Marshack stresses the differences between cave art and the carvings on the bones, he implies that cave designs might also have a seasonal meaning and have been executed at certain times of the year.

To believe fully in Marshack's vivid reconstruction of the Paleolithic mentality, we need more evidence—more evidence that the notations are truly lunar and not random, and more evidence that other theories do not explain certain animal designs better than a specifically seasonal one. (Could not pictures of young animals refer to general ideas of birth and death rather than to the coming of spring?) Marshack's work has certainly succeeded in drawing attention to the variety present in mobile art; variety in the style of the abstract notations adopted over many thousands of years, and variety in the realistic animal themes which in the later periods appear alongside them. Marshack often conveys the impression that all this represents one tradition, one way of thought or a coherent system of belief and practice throughout Europe. A perceptive critic has asked:

> Does the material reveal any more than the fact that man during this time was capable of cumulative symbolic recording of objects or events or phenomena, that he had considerable inclination to do so, and that he still did it in a varied and idiosyncratic manner?[7]

Marshack is at his most convincing not as an interpreter of Paleo-lithic philosophies but as an acute observer of individual objects. One of his outstanding contributions is an examination of over a dozen engraved bones from a site on the east coast of Italy called Polesini, connected neither with notations nor with seasonal animals. His study reopens the concept of hunting magic as it would apply conventionally to the wounded animals engraved on the Polesini bones, and shows how inadequate such a concept is to explain the complex nature of the carvings.

For example, several of the Polesini bones show parts of horses or unidentifiable animals carved in an indistinct, sketchy style, and not just once but repeated over again with different engraving tools (on one bone a horse has three consecutive muzzles, and on another a deer has four). These fragments of creatures are shown pierced with feathered darts of all shapes and sizes, which are sometimes reduced to almost unrecognizable abstract lines and are the product of many different toolpoints. So the Polesini animals do not seem to be care-fully observed seasonal images, nor are they simple magic figures sketched for immediate use only, but, instead, the act of renewing both the animals and the weapons on different occasions seems to have been important.

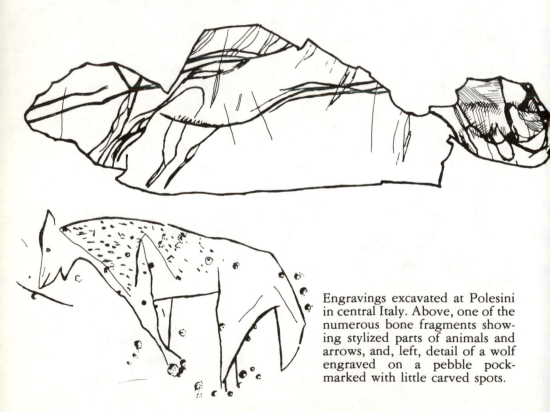

Engravings excavated at Polesini in central Italy. Above, one of the numerous bone fragments show-ing stylized parts of animals and arrows, and, left, detail of a wolf engraved on a pebble pock-marked with little carved spots.

Certain cave paintings and engravings may have been renewed in a similar fashion from time to time, although as yet Marshack has presented only a few such examples. One is a tiny "sanctuary" inside the decorated cave of La Pileta in southeast Spain, in which it is impossible to stand upright. This cavity has clay walls that are covered with complex red-and-black designs of animals, abstract arcs and rectangles, and weaving, serpentlike lines. Marshack notes that every available surface inside the chamber, including the far corners where there are no images, appears to have been lightly and repeatedly touched, as if there was some ritual act of fingering a "holy wall." Marshack has also examined signs consisting of short parallel lines found on or inside the animal outlines and usually described as "wounds" by those who favor a hunting magic interpretation. Each of these sets of double marks appears to have been made with a

View inside the chamber studied by Marshack at La Pileta, southern Spain, showing a horse with "wounds" painted in a variety of pigments.

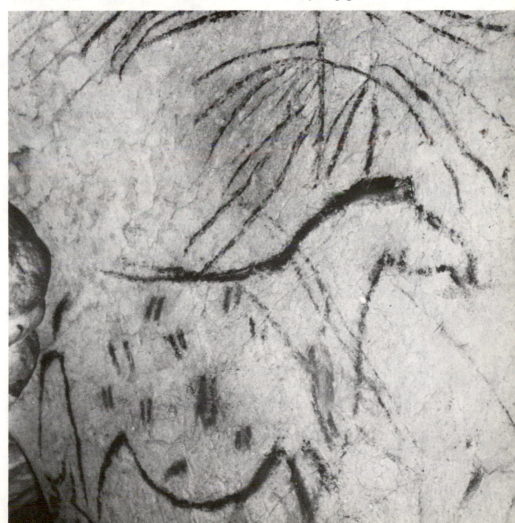

The multiple outlines of a reindeer engraved on the cave wall at Ste. Eulalie in the Lot valley, southwest France.

different pigment and by a different pair of fingers, resulting in signs of a varying color, size and spacing. The use of this cramped chamber for a rite connected with the animal images, involving the participation of several people over a certain period of time, seems likely.

Marshack's theories of the renewal of images, of repetitive, symbolic acts of carving, have not passed unchallenged. One of the most interesting alternatives was advanced by Michel Lorblanchet in trying to explain the multiple outlines that entangle the legs and heads of animals carved on the cave walls at Ste. Eulalie in the Lot valley. The realism of these animals, observed with superb accuracy in many different natural poses such as grazing and galloping, is curiously at odds with their superfluous limbs and multiple muzzles. So we encounter one reindeer with seven back legs, and a headless horse and an ibex with four or five front legs each. Eyes, ears and muzzles seem to have been renewed on various depictions of reindeer and horses. Although there is good evidence that many centuries separate the reindeer of Ste. Eulalie from those on the later Magdalenian stone blocks of Limeuil, nevertheless, the carvings at both sites share the same vivacity and the same extraordinary multiplication of body parts.

The most intriguing observation made by Lorblanchet is that the

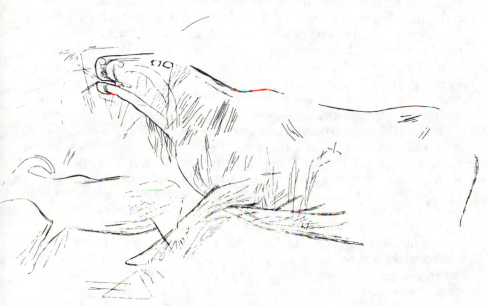

Another reindeer from Ste. Eulalie.

multiple heads and legs are shown not in parallel but as if they were all moving from the same shoulder joint. Why not, Lorblanchet asks (rejecting the "art school" theory and the "symbolic renewal" theory), consider these multiple outlines as an attempt to show movement, "a kind of pre-literate cinema"? The outlines always occur on animals with lively postures that already indicate action, while one final outline is always accentuated at the expense of others to avoid a monstrous "centipede" effect. In other words, *some* cases of repeated limbs or muzzles might represent no more than an artistic convention rather than the evidence of a symbolic practice.

There will always be differences of opinion in the detailed interpretation of such curious and complicated designs. The most valuable aspect of Marshack's recent work is his dedication to exploring certain types of marking that were ignored or neglected by many previous researchers and for which no proper explanations of any kind have been offered before.

For example, there are peculiar patterns of ribbonlike lines, sometimes called "meanders" or "macaronis," that twist and weave their way across clay or limestone surfaces in caves, or else appear on stone blocks excavated from a small number of living sites. Marshack is the first to study these serpentine patterns as a class of marking in its own right. The lines *do* appear on their own, but also alongside animal images, as at La Pileta.

He has defined some interesting features of this "tradition," if that is the right word to refer to so flexible and free a form of marking. The meandering lines were often extended bit-by-bit with the addition of extra lengths or side branches, while different people appear to have added to the patterns at different times. At Gargas in the Pyrenees the ribbonlike lines were engraved on the cave surfaces with flint tools, and they were also traced in soft clay with fingers and with a stiff brush, perhaps made of reeds or bristles. Anarchic "macaronis" outlined on a large clay ceiling in the cave of Rouffignac were clearly the work of a number of people who had differently sized and spaced fingers. The participation of several people at widely separated sites might suggest something more than mere childish "finger-painting" with which this activity is usually compared, although we may be skeptical of Marshack's idea that the meanders in some way express a water symbolism.

It is obvious that Ice Age Europeans expressed their beliefs and observations in many different ways. At some sites we have seen that separate engravings were piled on top of one another on stone and bone to the point of illegibility. Elsewhere, individual images were the subject of careful additions and renewals, sometimes for "cinematic" effect and sometimes, doubtless, with ritual intentions. Animals like the Montgaudier salmon were depicted in remarkable detail so as to suggest a reference to a particular time of year, but they could also be shown so sketchily (like the Polesini horses) that they seem to tell us only of an animal species in its most general, abstract sense. Animal carvings appear on their own, just as the notational markings can also appear isolated from any other type of carving. On a small number of objects dating from late in the Upper Paleolithic, the animals and the notations are combined in ways that suggest one or perhaps several special functions such as Marshack envisages. His pioneering work has been largely responsible for opening our eyes to this remarkable diversity of carved animals and signs, even though we may disagree with some of his conclusions. No single theory is likely to fit such a complex range of objects and imagery. Nor shall we easily resolve the mystery of what passed through the minds of Ice Age people as they worked at their strange "sketchbooks" by the rock face overlooking the grassy slopes of La Marche fourteen thousand years ago.

13: When the Forests Came

The Dance of the Birdmen

There are few better ways to experience the contrast between the Ice Age and modern times than to travel to Sicily during the summer months. The town of Trapani is an unlikely starting point for a Sicilian adventure, although until recently it was one of the chief Mafia strongholds in the island. Now it is a characterless industrial port, situated along the flat western coastal strip bordering the Mediterranean. At the dusty waterfront the traveler boards a hydrofoil ferry, which speeds over brilliant blue water toward a little shelf of bare limestone rock that interrupts the horizon. This is Levanzo, one of the tiny islands in the Egadi group, which lie about fifteen kilometers from the Sicilian coast. From far out at sea one notices that the honey-colored limestone of the Levanzo cliffs is pitted with the dark hollows of cave entrances.

After landing at the only village on the island, the traveler continues on mule-back along a solitary path which winds up and over the arid valley slopes of the interior. There is little to draw the fishermen of the village into these barren expanses of gorse, except for the occasional rabbit hunt. Soon the mules pick their way down to the stony ledge which forms the western shore of Levanzo, and into the cool arch of a cave entrance overlooking the sea. The back of this shelter tapers to a narrow passage, leading the visitor into the chilly darkness of the cave itself.

Here the visitor's torchlight falls on dull cementlike walls covered in black paintings. These are almost certainly prehistoric, but bear no resemblance to the graceful and spirited art of the Ice Age. There are a few crude sketches of four-legged animals, little more than black blobs attached to legs like sticks. These animals are outnumbered by many curious patterns which could be human forms, although some are so far removed from reality as to resemble insects or beetles.

Below these quick black sketches, lying almost at the feet of the visitor on sloping masses of stalagmite, are engravings of a very different quality. These are so delicate as to be nearly invisible, except when a torch is held to one side of the panel. Then one can follow the shapes of animals, traced out with one pure line for the body, uncluttered by details such as depictions of fur or muscle. The engravings are deceptively simple, for the single line flows with an assurance and naturalism utterly different from the black daubings

View from the cave porch at Levanzo, one of the Egadi islands near Sicily. Here a group of hunters lived towards the end of the Ice Age about 11,000 years ago.

above. Indeed, the quality of the engraved silhouettes is such that they could hardly belong to any age other than the Paleolithic. Here the paradox of Levanzo asserts itself, for one can step away from these fine tracings of wild asses, horses, cattle and deer out into the parched wilderness of an island where such animals could not possibly survive today. The sense of a vanished world is particularly strong.

The carvings at Levanzo are among the latest surviving examples of Paleolithic art. A limestone block engraved with an ox in an identical style to those inside the cave was unearthed from excavations under the cave porch. The layer from which the block came has been dated to about 9000 B.C., when the final centuries of the Magdalenian era were drawing to a close in western Europe. Yet the Sicilian hunters lay far beyond the influences of the Magdalenian world. The style of their art acquired little of the detail and intricacy of the French cave artists' work, while their day-to-day existence in the relatively mild conditions of the Mediterranean must have been very different.

Instead of following huge herds of reindeer, the Sicilians chased a range of animals similar to those depicted in the Levanzo engravings: mainly wild ass, oxen and deer. The remains of these beasts have been found in numerous cave sites occupied during the late Ice Age along the northern coast of Sicily, as well as at the entrance of Levanzo itself. The explanation for the presence of these animals on a remote island lies in the shallow waters which today are traversed by the hydrofoil from Trapani. At the end of the Ice Age, when the sea levels were lower by some thirty meters, the Egadi islands probably formed the rocky western promontories of a continuous land bridge, which not only joined them to Sicily, but possibly linked Sicily to the toe of Italy as well. No doubt the favored hunting grounds of the Levanzo hunters now lie under the Mediterranean.

Were the Levanzo engravers fired by ideas similar to the Magdalenian cave artists, even if their styles were so different? Unlike those of France and Spain, there are few well-defined abstract signs or symbols in the Sicilian caves. Another peculiarity is the presence of human figures which do not find a convincing parallel anywhere else in Europe. Levanzo has at least three of these strange figures, engraved with a fine line similar to the animals, and grouped together in a curious scene. Two of the figures have featureless fan-shaped heads, while the large central character with his straggly beard seems to be gripping the arm of a birdheaded person or creature to his right. A reclining fourth figure is painted in a conspicuous position at the back of the cave. It is the only red painting visible at Levanzo, and it, too, has a fan-shaped head, so it may well be Paleolithic in date along with the other engraved figures. It seems likely that the crude black paintings, mainly stylized human figures of a quite different character, were the work of the Neolithic people whose food remains and pottery were also found in the cave mouth.

The most remarkable development of this late Mediterranean art is represented at another Sicilian cave, this time situated on the mainland at Addaura. Here human figures not only occupy the center of attention, but their arrangement suggests a realistic sense of composition or even perspective lacking from almost all Paleolithic scenes.

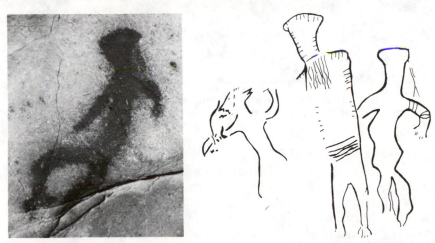

The red-painted figure at Levanzo and *(right)* engraved humanlike beings from another part of the same cave.

The central part of the engraved scene at Addaura, a cave on Monte Pellegrino, overlooking Palermo in western Sicily.

The decorated panel is located in a small cavity, completely illuminated by daylight and perched high in a steep overhanging cliff that faces toward the sea. The cliff forms part of Monte Pellegrino, the mountain which dominates the northern suburbs of Sicily's capital, Palermo. After the last war it was used for storing ammunition, which accidentally blew up. The explosion dislodged thick concretions that had entirely covered the prehistoric engravings, and revealed an extraordinary composition.

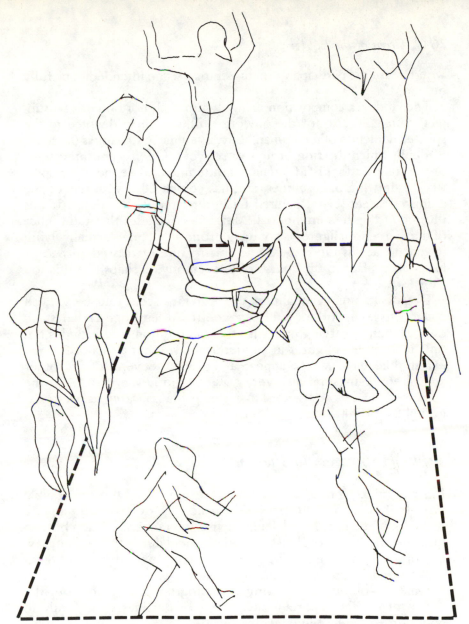

The central ring of engraved figures at Addaura. The perspective base line superimposed by Mezzena is shown. The reader is invited to draw his or her own lines and conclusions.

Above a series of lightly carved animal designs there appeared over a dozen well-proportioned human figures, depicted in a variety of lively attitudes. A group of them seem to be standing, leaping or waving their arms in a ring around two contorted figures, whose penises are clearly indicated. The awkward and, indeed, painful positions of the central pair may possibly be explained by the lines which run close to their backs: these could represent ropes. Several of the faces seem to end in beaks rather than noses, so we may wonder: are the participants in this strange scene intended to be fully human?

The Addaura composition is highly unusual, not only for its subject matter but also for the obvious skill and effort devoted to the spirited depiction of the human body. Even more striking is the sense of depth which the ring of figures conveys. One recent interpreter of the composition, F. Mezzena, has attempted to define its perspective by drawing in base-lines, and has compared it with one of the animated street scenes painted by Peter Breugel the Elder. If this illusion of depth is intentional, Mezzena concludes that the higher of the two central figures is not lying prone at all, but is actually flying in mid-air across the ring. Could this tossing of trussed-up people have taken place in the course of a ceremony, perhaps an initiation rite?

Despite its unique qualities, the Addaura scene can be viewed confidently as a work of late Ice Age art. At least one earlier stage of much fainter carving underlies the composition, and includes not only more human figures in an identical style, but also a deer head exactly like those which appear at Levanzo. Several of the other animal engravings are also very close to the Levanzo style, so the combined evidence suggests that the Addaura carvings were executed about 9000 B.C.

What Happened to the Cave Artists?

It is natural to wonder what could have stimulated this remarkable growth of vision at Addaura. There is no such development among the work of the final Magdalenian artists in France and Spain but, on the contrary, there are signs of stagnation and decay. Animal engravings of the highest quality *are* known from this period, some of them —like the browsing and galloping reindeer carved on stone blocks at Limeuil—observed with almost photographic accuracy. But the output of certain important sites stops, or declines significantly in both number and quality, at this time.

An interesting case is La Vache in the central Pyrenees, a cave with an occupation deposit that has yielded some of the most beautiful of all Magdalenian engraved bones and antlers. A careful study of the successive layers in the deposit and the artworks that came from them has revealed a spectacular decline in naturalistic subjects from the final layers of the site. The excavator concluded that

> the last quarter of the eleventh millennium marks a break, a fundamental aesthetic divide, in the mobile art of the cave of La Vache.[1]

So, at the end of the Ice Age, came a gradual faltering and dimming of the vision which had sustained the cave artists for so many thousands of years. At some undefined point around 9000 B.C., it was all over: realistic animal carvings were no longer produced by the hunt-

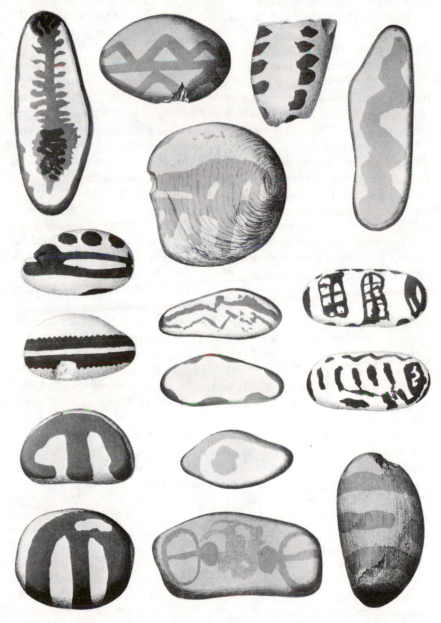

A selection of painted pebbles from Le Mas d'Azil unearthed by Piette's diggers.

ers of France and Spain. In their place we have a few scattered examples of a totally abstract art form, which make a depressing comparison beside the achievements of the Magdalenians.

These simple designs consist of a variety of chevrons, dots, parallel lines and S shapes painted in red ocher (or, more rarely, engraved) on the faces of little river pebbles. At the turn of the century, the laborers who dug for Édouard Piette discovered a large number of the red-painted pebbles at the tunnel cave of Le Mas d'Azil in the Pyrenees. Piette subsequently learned that a few pebbles had been deliberately tossed in the gutter and in the river, while many more had been missed and lay in the spoil heaps. Despite the carelessness of his work force, Piette was able to determine that they had come from the layer which succeeded the last Magdalenian remains, and, as no comparable finds were then known elsewhere, Piette gave the name "Azilian" to the pebbles and the tools discovered in this layer. In recent times, Azilian tools have been found in many regions settled by the late Magdalenians, while painted pebbles just like those from Mas d'Azil have been excavated at over half a dozen sites.

With his characteristic enthusiasm, Piette advanced the idea that the patterns represented primitive writing, an alphabet of pebbles, or else a system of numbers—a theory which cast the Azilians in the role of pioneers of civilization. However, the variety of the designs suggests that they are more plausibly interpreted as tallies or counters, or else as symbols of a less precise kind connected with ritual ideas. A few complicated engravings may even be notations like those studied by Alexander Marshack, and he has, indeed, proposed a lunar explanation for the intricate notches on one Azilian bone. It is conceivable that the practical aspects of recording time or the seasons were still of importance to the hunters, even when the values connected with painting and engraving animals had lost their meaning.

Yet most of the neatly painted pebbles look very different from the carved stones dug up from Magdalenian layers in France, with their chaotic tangles of animal designs. To find anything like these abstract, formal patterns, we must look at sites in the Mediterranean, especially along the coastlines of Spain and Italy. At Parpallo and Romanelli, blocks carved with parallel lines, "ladders," chevrons, and meandering "serpents" have been found in considerable numbers. They are particularly common in the Italian sites, appearing alongside rather stiff and simplified animal outlines. These geometric designs were apparently never adopted further south by the Sicilians; they formed no part of the repertoire of the artists at Levanzo nor at Addaura, where the unique scene of human figures was carved about 9000 B.C. But, at about this time, the influence of the Mediterranean style may somehow have reached the hunters of the Pyrenees, and here the painted patterns replaced the last vestiges of Magdalenian art. Alternatively, the Azilians may have invented the simple abstract designs for themselves, without any help from the Mediterranean world.

For a relatively brief time, perhaps for a thousand years or so, this rudimentary art persisted both in southern France and Italy, and then it vanished altogether. After about 8000 B.C., in many regions of Europe that were now transformed by changes in climate and vegetation at the end of the Ice Age, the hunters had no art at all. Although communities in these regions survived by hunting and gathering as before, some aspect of the postglacial world discouraged all major forms of artistic expression that could have survived until modern times.

The full story of prehistoric art does not, of course, end here. The great achievements of the Neolithic and Bronze Age communities were yet to come, notably the monumental architecture of temples and palaces such as Stonehenge and Knossos. Yet during the four or five thousand years which elapsed between the end of the Ice Age and the slow spread of a settled farming existence throughout Europe, there was no widespread artistic revival. We know that during this period at least one visit was paid to the galleries of Niaux, and perhaps to other decorated caves as well, yet whether the ancient painted designs were comprehensible or even noticed by the explorers we cannot say. In the few regions where painting and engraving *did* appear during the period 8000 to 4000 B.C., the influence of the Paleolithic traditions seems to have been slight. Much of this postglacial art is so different in quality and spirit from that of the Magdalenian era that it is hard to believe old traditions lingered on in the artists' minds.

Along the coastal plain and highlands of eastern Spain, there are dozens of rock shelters decorated with lively hunting scenes, mostly executed in red ocher. Although these paintings are usually situated on the back walls of south-facing overhangs fully exposed to the intense summer sunlight, they have nevertheless survived from a remote prehistoric period. They display a development of artistic vision comparable in certain respects to the exceptional work of the engraver at Addaura. A fluid sense of composition and action is often present, while isolated and idealized animal forms are rare. Human figures occupy the center of the stage in many of the scenes. Their attenuated limbs—muscular legs flailing in hot pursuit of galloping boar and ibex—convey a sense of hectic activity. In many cases the activity appears to be entirely secular in character: the paintings look like commemorations of hunting exploits or of skirmishes between opposed groups of bowmen. Most of the hunters appear to be naked, although certain figures display elaborate headdresses, one-piece skirts (on both male and female subjects), or short trousers similar to a traditional garment still worn by country people in the Aragon region.

None of these lively scenes has ever been found deep inside a cave in eastern Spain. Nevertheless some researchers, including the Abbé Breuil, believed that the rock shelters were decorated during the Ice Age by hunters who had developed their own peculiar customs of painting, uninfluenced by contemporary Magdalenian cave artists in France and Spain. Breuil thought that the rare depiction of certain animals which he regarded as typical of the glacial period (notably elk) supported his case. However, the most common animals shown —deer, goats, ibexes and boars—are exactly the species which would have been best suited to the relatively mild conditions of postglacial times, and to the parklike landscape of the region at this period (even elks are most at home in lightly wooded surroundings). The size of the red-painted figures is invariably small; animals only a few centimeters across are common, unlike the monumental scale of many works of cave art. Furthermore, no Paleolithic tools have ever been discovered at any of the eastern Spanish sites, but instead industries

Part of the great battle scene painted at Valltorta in eastern Spain. Such obvious depictions of social conflict are absent from Ice Age art.

of a post-Ice Age type have been found at richly ornamented shelters such as those at Cogul and Albarracín.

The exact dating of the most accomplished examples of this art remains uncertain. Despite the absence of agricultural scenes of ploughing or herding from the shelters (although a bowman seems to be leading a goat by a rope at one site and a horse at another), some recent evidence suggests a date surprisingly close to the time when cereal crops were first grown in the western Mediterranean, and when the herding of sheep and goats was probably already widespread in favorable regions. Could it be that the exploits of hunters had gradually acquired a heroic value at exactly the period when the hunting way of life was on the wane? In any case, if it is true that the bulk of these vivid compositions are from the period around 5000 B.C. and later, then their independence from the Ice Age world is clearer than ever.

In general, then, the achievements of the Paleolithic cave artists of France and Spain seem to have passed away without issue. Whether Ice Age traditions from other regions of Europe, such as Moravia and the Ukraine, continued to inspire the craftsmen of the

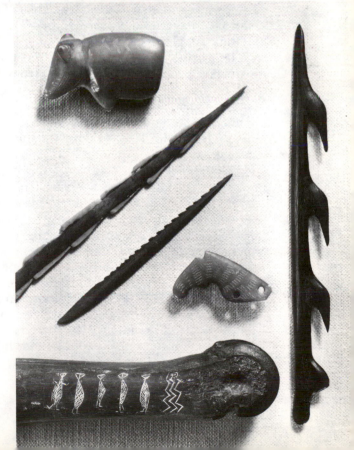

Harpoons and other objects fashioned in the Maglemose art style from Denmark.

north is a more difficult question. The tools and weapons of the postglacial hunters are usually undecorated, except in the regions of southern Scandinavia around the shores of the Baltic, stretching into northern Germany and parts of Poland. Here the hunting peoples of the "Maglemose" culture adopted a distinctive type of ornament, created by patterns of tiny holes drilled into bone and antler surfaces. These patterns often took the form of intricate, repetitive networks of little triangles, chevrons and lozenges, occasionally placed next to extremely stiff and stylized "stickmen" or clumsy animals.

The regimented designs are a contrast to the freedom of inspiration of many engraved Magdalenian objects, yet there are abstract patterns quite close to the Maglemose ones known from Ice Age sites in Russia. It is also intriguing that this formal type of decoration was often applied to the surfaces of highly polished antler staves pierced with a hole, identical to the batons which seem to have been such a widespread, potent symbol during the Ice Age. Finally, the Maglemose artists also worked delightful animal sculptures from small lumps of amber and other materials. Alexander Marshack has pointed out that the body of at least one of these sculptures, a little amber bear, is apparently covered with rows of successive notation marks and shows signs of long-term handling, like a variety of other objects from Ice Age sites which he has carefully examined. So, once again, the substance and spirit of Paleolithic art seems to have foundered, while certain rudimentary habits of decoration or observation may have lingered on, perhaps for a substantial period, among the hunting peoples of the Baltic.

The passing of Magdalenian art has profoundly affected attitudes to the postglacial period. If we turn from the best Paleolithic engravings to the stickmen of the Maglemose or to the painted dots and lines of the Azilians, it certainly seems as if the creative faculties of hunters at the end of the Ice Age were fatally sapped and undermined. It is difficult to write of this period without suggesting that some great collapse—whether of a spiritual, economic or social kind —took place around 9000 B.C. Distinguished prehistorians have occasionally written as if a "crash" of unprecedented proportions severed the brilliant world of the Ice Age artists from that of their impoverished successors. One expert, for instance, has referred to the "catastrophe" represented by the "collapse" of reindeer economies in western Europe.[2] Another mentions that the postglacial hunters lacked the "brilliance" of the Upper Paleolithic people, who were living in "the stimulating landscape of the Ice Age,"[3] as if their successors, hemmed in by the gloomy forests, were simply too depressed or bored to express themselves. In trying to describe changes which occurred gradually over the course of century after century, and which are still poorly understood, it is always tempting to simplify and speak of "crashes" or "catastrophes."

Whatever the exact circumstances were that discouraged the production of artworks, there are no obvious disasters or breaks apparent in the archaeological record. The majority of the flint and bone tools manufactured by the Azilians from about 9000 B.C. are indistinguishable from those made by the Magdalenians. Furthermore, there is clear evidence to show that in certain areas the late Magdalenian hunters were contemporaries of the early Azilians, so there was no uniform break with the past.

In other cases, at Duruthy in the western Pyrenees for example, the old living bases of the Magdalenians were deserted as the oak forest closed in. After a break, Duruthy was again occupied, this time by the Azilians, as one last cold snap reopened the landscape and brought the reindeer back briefly to the Basque country. Despite this return to a chilly parklike landscape, conditions were probably less favorable for big-game hunting than in Magdalenian times. The Azilians were so few in number that they occupied only a small fraction of the area of Duruthy, and it was not long before the reindeer became scarce and they were forced to seek red deer instead. While these marked changes in their surroundings and way of life were in progress, the Azilians seem to have continued to make a tool kit similar in all essentials to that of the late Magdalenians. Eventually, the forest dominated the foothills once more, and Duruthy was finally abandoned after serving as a shelter for perhaps as many as one thousand generations of Upper Paleolithic hunters. Viewed in such a perspective, the art and tools of the Azilians appear like a last, feeble ghost of the great Ice Age cultures which had preceded them.

It would probably be a mistake to imagine that the disappearance of the reindeer posed a food crisis which seriously threatened the survival of the hunters. Yet the economic shift to animals with less extreme seasonal movements, like red deer, elk and ibex, may have had important social consequences. The ancient network of farflung contacts, which had been sustained partly by the practical necessities of migration, must have diminished steadily in importance. This could explain why the immemorial values and relationships expressed in Paleolithic art slowly declined, while the everyday lives of the hunters may actually have become more settled and secure.

Forest Fires and Seashells

One fact which recent research has made abundantly clear is that the postglacial hunters were not passive, timid creatures of the forest, but met the challenges of a changing landscape with all the resourcefulness of their forebears.

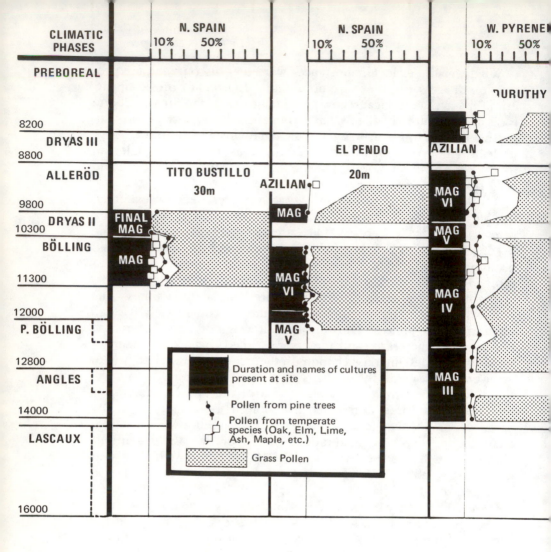

CLIMATIC PHASES	N. SPAIN 10% 50%	N. SPAIN 10% 50%	W. PYRENE 10% 50%
PREBOREAL			
			DURUTHY
8200			AZILIAN
DRYAS III		EL PENDO	
8800			
ALLERÖD	TITO BUSTILLO 30m	AZILIAN 20m	MAG VI
9800		MAG	
DRYAS II	FINAL MAG		MAG V
10300			
BÖLLING	MAG	MAG VI	MAG IV
11300			
12000		MAG V	
P. BÖLLING			
12800			MAG III
ANGLES			
14000			
LASCAUX			
16000			

Legend:
- ■ Duration and names of cultures present at site
- ● Pollen from pine trees
- □ Pollen from temperate species (Oak, Elm, Lime, Ash, Maple, etc.)
- (dotted) Grass Pollen

This chart simplifies the environmental evidence at a number of key sites in France and Spain during the closing stages of the Ice Age. The pollen record of grass, pine and of less hardy trees gives some idea of the responses of vegetation to the changing conditions. The ordering is based on radiocarbon dates and on the climatic evidence itself. Notice the interesting time overlap of the different cultural styles. At the bottom, two reconstructions of extreme environments: Europe during the most severe phase of the last Ice Age and *(far right)* during the mild conditions of the Alleröd interstadial.

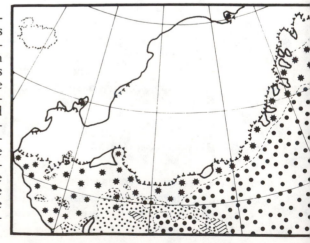

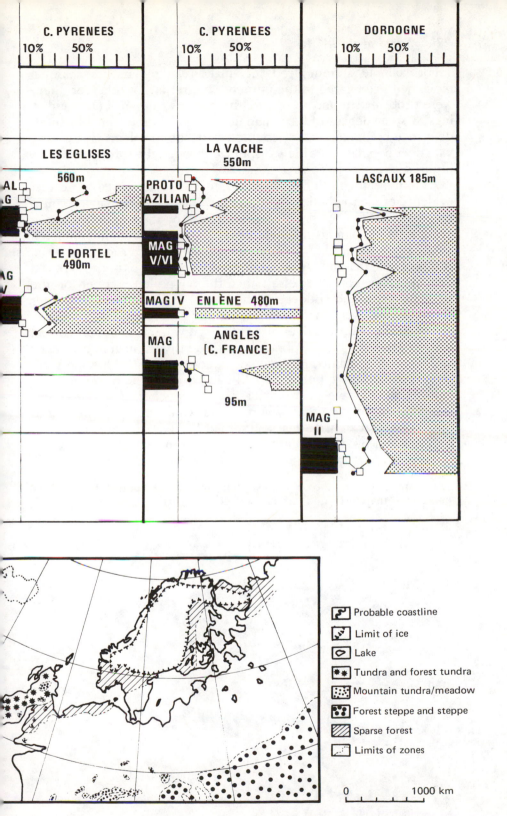

C. PYRENEES	C. PYRENEES	DORDOGNE
10% 50%	10% 50%	10% 50%

LES EGLISES

560 m

AL G

LA VACHE
550m

PROTO
AZILIAN

LASCAUX 185m

LE PORTEL
490m

AG V

MAG
V/VI

MAGIV ENLÈNE 480m

MAG
III

ANGLES
[C. FRANCE]

95m

MAG
II

Probable coastline

Limit of ice

Lake

Tundra and forest tundra

Mountain tundra/meadow

Forest steppe and steppe

Sparse forest

Limits of zones

0   1000 km

For example, although red deer make relatively modest seasonal movements compared to the immense migrations of reindeer, there were probably still incentives for the hunters to limit their wanderings by some degree of herd management. The ease with which red deer may be domesticated, coupled with the problems of tracking wild deer populations widely dispersed through the forest, may have inspired attempts to control them.

As in the case of reindeer herding, it is difficult to establish direct evidence of such activities, but there are one or two tantalizing clues. Large amounts of ivy have been identified by their pollen traces at postglacial settlements in England and in the Netherlands, situated where ivy is never likely to have grown naturally in such profusion. It is probable that the ivy was gathered by hunters either to lure red deer into an ambush, or else (since the ivy pollen is found on the settlement sites and not deep in the forest) to provide winter fodder for herds that were kept under some form of control.

The image of the unadventurous forest dweller is so deeply rooted that only recently has the possibility of extensive human interference with the natural landscape been considered seriously. One spectacular way in which the hunters could have affected the movements of animals would have been to set fire to stretches of forest cover, especially in upland areas, where game would be flushed out into open country. Extensive charcoal remains forming a fire horizon can be identified by digging into soil dating from postglacial times or

A century-old print showing Australian Aborigine hunters flushing out game with the aid of deliberately started bush fires.

even earlier. Indeed, the widespread occurrence of these horizons during one of the warm intervals close to the end of the Ice Age suggests that hunters in southern England and the Low Countries were already active in their efforts to "ransack" forested areas.

Of course, many of these fires may have started accidentally, but there were some very obvious advantages to be gained by regularly opening up the forest by fire. New clearings could support far greater numbers of animals than the forest cover itself, and indeed animals would be attracted from a wide area. When deer populations browse in freshly created clearings to feed on the regenerating shoots, they display phenomenal, if fairly temporary, increases in their body weight, breeding rate and over-all health. It is scarcely likely that skilled hunters would fail to observe these facts.

The evidence for such an exploitation of the forest comes from several English sites, where charcoal layers are closely associated with discoveries of flint tools. In fact, a deliberate policy of burning may been pursued on a more widespread scale than these scattered finds would suggest. At two separate intervals during the postglacial era, grassland and fire-resistant hazel bushes formed the chief vegetation cover in parts of southern England and Yorkshire. While the interpretation of pollen evidence is always a delicate matter, some environmental experts doubt that natural climatic factors could be entirely responsible for this type of landscape.

If burning was really being practiced on such a large scale, then the post-Ice Age hunters may have initiated serious long-term damage to their surroundings, which they could scarcely have foreseen. It is possible that burning of the high margins of the forest could have set off the processes of soil deterioration and blanket peat formation which created the wastelands of Dartmoor and the Yorkshire moors. The transformation by academics of the dependent hunter into an environmental vandal is rather startling.

The postglacial hunters of the north were also pioneers in a positive sense, for they colonized many regions which had previously been covered with ice or were uninhabitable because of extreme cold. For example, the earliest signs of settlement in Ireland have been uncovered at Mount Sandel in Ulster, where a substantial oval wooden house and other remains have been found, dating to a period not long after 7000 B.C. In succeeding centuries, small groups extended the frontiers of settlement along the coastlines of east and west Scotland, including the Hebridean islands, and along the shores of the Baltic.

There are a number of factors which could explain this activity besides a pioneering urge on the part of the postglacial hunters. The increasing numbers of settlement sites in most parts of northern Europe suggest a steady rise in population throughout the period. Pressure of numbers (and, later, competition from early farmers)

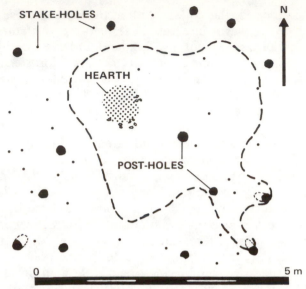

A plan of the earliest house in England recently excavated at Broom Hill, Hampshire. This structure with its sunken floor was built by a group of post-glacial hunters in about 6500 B.C., while the site itself was settled for at least 2,000 years afterwards.

could have pushed the hunters steadily northward. Another crucial factor was the development of sturdy oceangoing boats and deep-sea fishing, which were probably indispensable for a secure livelihood among the lochs and fjords. Rock carvings from northern Norway, which many consider to belong to a period when farming was already widely practiced in more favorable regions of Europe, not only depict elk and deer, but also line-fishing from small boats which look like the rugged skin-covered Irish curraghs or Eskimo kayaks of recent times.

The extent to which there was a shift to marine resources of all kinds throughout Europe at the end of the Ice Age is a tricky question. One popular theory, advanced by Lewis Binford and others, states that the forests essentially pushed the hunters out toward the coastlines, and it was here along the shore that the major centers of population developed. The reliability and richness of the resources there, especially shellfish and waterfowl, allowed a fairly settled existence and, in these sedentary conditions, human populations grew quickly. When these expanding communities impinged on inland neighbors, who were still following their old seasonal round, pressure built up both between and within groups to compete for resources and to find new methods of procuring food. Such social tensions eventually paved the way for the adoption of farming.

This rapid sketch of a theory may seem quite plausible, but the difficulty comes in finding detailed evidence to support it. We simply cannot establish a reliable picture of how intensively the coastlines were exploited either at the end of the Ice Age or in the succeeding postglacial period. During this time, the seas were still rising steadily, fed by the melting ice caps and glaciers, and destroying most of the earlier remains of coastal bases.

Thick rubbish heaps of shells dating to the postglacial *are* preserved in favorable spots along the present European shores. Substantial shell mounds have been uncovered in the Tagus estuary in Portugal, on the southern coast of Brittany, and on the islands of the Inner Hebrides. One early commentator attributed these remains to the invasion of a race of people with a new taste for limpets and winkles.

Modern research on the significance of the shell mounds is the subject of intense debate and disagreement. Many shellfish species are highly nutritious, and can be harvested easily at any season of the year; in theory, at least, a diet of mollusks could provide the basis for a settled existence. On the other hand, some populations of marine mollusks are sensitive to excessive disturbance and exploitation, and furthermore there was little inducement for the hunters to depend on this rather monotonous food source. Far better to visit the rocks and pools for a month or two during the lean season, and then move inland to hunt game such as deer and elk for the rest of the year. This is the pattern which has emerged from scientific study of the shell mounds of Denmark and the Hebrides. Would such a seasonal round have been very different from the way of life led by their Paleolithic ancestors?

Nevertheless, in more favorable areas of Europe, where the activities of shellfish gathering, sea fishing and hunting could be combined easily, year-round bases may have developed. Proof of the permanent nature of these sites is usually lacking, because the laborious task of examining fish bones and the growth rings on mollusk shells is all too rarely undertaken.

This is exactly the case in the Cantabrian region of northern Spain, where there are substantial shellfish deposits located in cave entrances. Some of the deposits, or *concheros* as they are known in Spain, are from Magdalenian times, but thick layers of shellfish remains are also dated to the period 7000 to 5000 B.C. The character of the hunters' lives at this time cannot have been very different from that of their late Ice Age forerunners. Red deer and ibex still remained important sources of meat, as they had been in the days of the cave painters at Altamira. Yet milder conditions may have made life easier: it is likely that a broader range of vegetable foods was available, and that the distances over which the deer roamed from season to season were more limited. Many of the *concheros* are ideally placed to be base camps for expeditions to estuaries, mountains and forested valleys, while at present no sites of this period are known from further inland. So the regular movements which had characterized the lives of their forebears may have become increasingly unnecessary for the people of the Cantabrian coast. A settled existence may have been possible with little extra effort, and with no major shift to new, previously neglected resources. At any rate, this is one likely version of events, for which conclusive proof is still lacking.

Yet even if the postglacial hunters were able to establish year-round bases in some regions of Europe, there is no reason why this development in itself should have led either to an inevitable population explosion or to the invention of farming. If a more secure livelihood resulted in an increased birthrate, then new risks, such as the harboring of disease in storage or waste pits, may have increased the death toll. It is also hard to see why social tensions would necessarily be resolved (as in Binford's theory) by the devising of new techniques for procuring food. An increase in population could be met in all sorts of ways, such as active attempts to control it by abortion and infanticide. More people could be supported simply by an increased workload, stimulated by some code of values which made the extra effort seem worth while.

The northwest coast of America was one region where dense populations were supported in permanent villages by a fishing and gathering economy. This was made possible not just by skill and inventiveness or by exceptionally rich resources, although certainly the Pacific coast Indians had these in abundance. Such factors were all involved in an intense social drive for prestige, acquired mainly through material exchanges and displays of wealth at great ceremonial feasts (the famous "potlatches"). The feasts were held by the hereditary chieftain and people of his local group, who played hosts to another related chief and his retinue. The events were arranged on the occasion of an individual's change of status, such as the marriage or coming of age of a high-ranking person. Occasionally they concerned rival claims to inherited social positions and became strongly competitive affairs, and then the gift-giving involved extraordinary quantities of material items (33,000 blankets were given away at one potlatch, while fifty seals were eaten at another).

The practical effects of the potlatch system—the way in which it may have encouraged extra work, or helped to balance out abundances in one area and shortages in another—are hotly debated by anthropologists. The important point is that agriculture was not necessary to support a level of population and social complexity far removed from the world of the desert hunters. If the wandering Australian Aborigines of the Western Desert with their few material possessions lie at one end of the hunting and gathering scale, then the settled northwest coast Indians, some of whom owned slaves and inherited hierarchical titles, represent the other end. Exactly where the hunters of the Ice Age and of the succeeding era fit in between these extremes is difficult to say. In any case, no single complication of the hunters' lives led inexorably to the abandonment of hunting and gathering. Fixed home bases and gradual increases of population could all have been sustained by the ancient way of life. The record of sites in postglacial Europe reveals just such a steady growth of settlement, not a sudden population crisis which would have seriously disrupted the social and economic order. Indeed, hunting and gathering has been the successful basis of such a wide range of societies that it is natural to wonder why agriculture ever arose in the first place.

The Fall

Faced by this question, some commentators have evaded the task of supplying a reasonable answer by retreating to an almost mystical outlook. It was "destiny," or "human nature," which led to the triumph of civilized culture over the barbarous instincts of the hunting past. The American prehistorian Braidwood explained why agriculture had not been invented at an earlier period simply by the claim that human culture had not been ready for it. In these terms, the transition to farming has the quality of an effortless miracle, which took place on the day when a flash of insight lit up the brain of a hunter and revealed the obvious advantages of agriculture.

These advantages, whether connected with the increased capacity for acquiring possessions, or for enjoying a more stable social life, are of course obvious in hindsight to the modern eye. To assume that prehistoric people would be spontaneously won over by the appeal of such ideas is to ignore the drawbacks of agriculture. High levels of production are not the product of miracles, but of hard human effort. Every increase in the output of an economic system has involved more work, not less, and there will be resistance to such a change, as the development agencies in Third World countries have experienced. The Hadza of Tanzania were one recent hunting people whose reluctance to take up the practices of their agricultural neighbors was directly based on the longer hours of work involved.

This is why a rational approach to the origin of farming demands that we try to identify the specific forces which eventually overcame the conservatism of the hunting way of life. There must be pressures which resulted not merely in high levels of population and social complexity (the northwest coast hunter-fishers certainly had these), but in the fundamental change to new methods of subsistence.

Did the change happen all at once, with the crisislike quality which popular terms like the "Neolithic Revolution" convey? Some of the major steps had already been taken long before the end of the Ice Age. We have seen that management of reindeer and red deer herds may well have taken place on a large scale, although the evidence of animal bones dug up from a site cannot by itself prove this. Dogs are likely to have been domesticated before the postglacial period, probably in several different regions of Europe and America. Mortars and grindstones, together with flints bearing the characteristic sheen or polish associated with the cutting of plant stems, are found at a number of late Paleolithic sites in Europe, and may reflect a steady growth in the importance of wild grasses and other vegetable foods in the diet. The interest of the Ice Age period lies not merely in the variety of relationships which the hunters established with their surroundings, but also the extent to which these activities foreshadowed things to come.

Indeed, it may be wondered if our everyday terms such as *hunting* and *farming, Paleolithic* and *Neolithic,* are not meaningless rigid distinctions drawn across a continuous process of change. Clearly there was no single technological invention or revolution, but a whole range of them which occurred in different places and at different times, some of them separated by many thousands of years. The problem of identifying important single factors that promoted the changes is more difficult than ever.

The coming of agriculture to the arid highlands of central Mexico is a good example of the complex processes at work. By about 7000 B.C., a progressive drying of the climate probably led to a decline in the wild animal population, and this may have been one factor that encouraged the hunters to experiment with the cultivation of plants. Some of the earliest recorded forms, however, are nonedible species such as the bottle gourd (used for containers), or comparative luxuries such as the avocado, pumpkin and chili peppers, all of which may have been cultivated by about 6000 B.C. Even maize, when it was eventually transformed into a recognizable domestic form by 5000 B.C., is unlikely to have been in any sense a staple crop. Its low yield at this stage (the early cobs were less than three centimeters long) probably ensured that it played a minor part in the economy for several thousand years.

Why hunting and the collection of nuts and legumes were eventually superseded by maize and bean cultivation remains uncertain. The fact that this occurred over several thousand years suggests that the maize-bean combination flourished under human control, and occupied a steadily more important part of the time and energies of the Mexican hunters in the course of their seasonal round. It is conceivable that in one sense the plants "took over" man, rather than the reverse. As one recent commentator has put it, "the long time span involved indicates that the Mexican hunter-gatherers were not pushed into agriculture at all, but drifted into it by accident."[4] In any case, simple social pressures cannot account for the early stages of crop experiments. Permanently occupied living sites are unknown in central Mexico before about 1500 B.C. In the arid highlands an extremely mobile hunting life was always necessary, and so dense populations or the demands of settled life cannot have figured in the pioneer efforts to cultivate maize.

In the Near East the circumstances were different. The end of the Ice Age brought changes to the vegetation and to the animal populations of a far less conspicuous kind than in western Europe. In the Fertile Crescent, which extended from Palestine and Syria around to the foothills of the Zagros Mountains in Iran, there was no blanketing spread of dense forest cover to disrupt the habitats of key game animals. In the Zagros, for example, the steppelike conditions of the Ice Age gave way to a savannah landscape dotted with oak and pistachio trees. These open conditions certainly favored the spread of wild grasses, among them the varieties of wild barley and wheat that were ancestral to modern forms. Yet the hunting or herding strategies of the Near Eastern peoples were remarkably undisturbed by the altered conditions. Recent excavations at sites such as Nahal Oren in Palestine have demonstrated a close continuity in the pattern of gazelle remains from the time of the late Paleolithic cultures until about 6000 B.C. At least one researcher claims that there was no significant difference in the way in which the people of the Zagros were exploiting sheep in Mousterian times and their Neolithic descendants in the same region over 40,000 years later. Either the transition from "wild" to "domestic" animals does not show up in the Near Eastern archaeological record, or else the essential steps toward the close supervision of sheep and goats had already been taken long before by the pioneering communities of the Ice Age. The relationship between man and beast seems to have remained essentially constant, at the same time as profound changes were taking place in other aspects of society.

There are several theories, none of them entirely satisfactory, for the beginning of agriculture in the Near East. The best known of these was advanced by the man who invented the "Neolithic Revolution," V. Gordon Childe. His work helped to fix attention on the Near East as the center from which the earliest farming techniques "radiated," and eventually spread to the rest of prehistoric Europe. Childe's explanation for the origin of agriculture in the Fertile Crescent emphasized climatic change rather than the effect of population increase or technological inventions. At the end of the Ice Age, Childe believed, a progressive drying of the Near Eastern climate led to difficult conditions for both humans and animals. The drought forced them to crowd together around the available water sources,

> that were becoming constantly isolated by desert tracts. Such enforced juxtaposition might provide that sort of symbiosis between man and beast implied by the word "domestication."[5]

Subsequently, little positive evidence was discovered to support the idea of a period of increasing drought, and Childe's theory fell into disfavor.

Nevertheless, as more precise information has come to light from new work at Near Eastern sites like Nahal Oren, Jericho, and Abu Hureyra, a degree of support for the old idea of a period of climatic stress has emerged. The impact of this climatic trend was probably not as dramatic as a term like "Neolithic Revolution" would imply. It may, however, have been sufficient to encourage a shift towards increasing dependence on plant foods.

A fundamental point is that the nature of the plant resources themselves may have been affected by climatic change, a "Botanical Revolution," without which the earliest cultivation in the Near East could not have succeeded on the scale that it did. It has been suggested that only with the arid high-temperature conditions of the postglacial period could the *annual* forms of cereals and legumes flourish, and only these forms could provide an efficient basis for agriculture. The laborious effort involved in collecting the seeds of perennial grasses would have ensured that plant gathering remained a relatively small-scale activity, but it was the perennial forms which were least favored in times of drought.

The exact circumstances of the human revolution which followed this botanical change are still the subject of intense controversy. The latest findings of excavators and specialists appear to indicate a vital shift in the pattern of settlement just before 7000 B.C., which may correspond to a period of stress, as Childe envisaged. Before that date, occupation in the upland caves and settlements of Palestine and Syria seems to have been continuous, stretching far back into the Ice Age. The gazelle was the most important animal throughout this period, and it may well have been herded exactly as sheep and goats were at a later date. The rugged terrain in which most of the early sites are located suggests that the potential for cultivating cereals was low.

By 7000 B.C., while occupation still continued for a while in the upland sites, the first signs of settlement in the fertile lowland plains appear at the great tells, such as Mureybit, Abu Hureyra, and the forerunner of the Biblical city of Jericho. Here the capacity for arable cultivation was great, and the settlements soon expanded to a remarkable scale. Not long after 7000 B.C., Jericho is estimated to have had two or three thousand inhabitants, who lived in substantial clay-brick houses. They had no knowledge of or use for pottery, and they continued to tend their gazelle herds just as before. Although the domestic forms of wheat and barley were not widely established at these sites until much later (around 6000 B.C.), it is scarcely believable that settlements of this size could have been supported without agriculture. So the earliest experiments with cereals of wild forms may well date back to the beginnings of occupation at the fertile lowland sites. Sheep and goats replaced gazelles for a number of reasons, an important factor probably being the capacity of sheep and goats for giving milk.

Although the evidence is always more complicated than a brief summary can convey, cereal cultivation in the Near East seems to have happened because of the interplay of a number of factors with a changing environment. These factors—the spread of annual grasses, a human shift to sites on good arable land, and steady increases in population—must all have reinforced one another, and have progressively diminished the importance of hunting and gathering.

The old way of life was not doomed because it was an inadequate basis for coping with survival in a changing landscape. Nor, as we have seen, was the coming of agriculture always the product of a conscious, rational design, an inevitable consequence of human progress. However, whether by accident or choice, once the Near Eastern communities had established arable farming on a widespread scale, the pressures on neighboring hunting peoples must have steadily mounted.

At present there is still little evidence to indicate what these pressures may have been: whether there was a substantial migration of people from the Near East, or whether existing hunting communities were gradually assimilated into the new way of life. It is easy to picture this process, like that of the fate of the Neanderthals, in the violent imperialistic terms of the recent colonial past, to imagine traditional hunting territories disrupted by pasture land, or their inhabitants pressed into the service of alien chiefs eager for power and prestige. The earliest farmers of eastern Europe, whose settlements date back to at least 5500 B.C., certainly manufactured a very different range of tools, and settled in quite different locations, from their hunting forebears. In contrast, the inhabitants of northern Italy and southern France gradually acquired the techniques of pottery-making, cereal cultivation, and herding, yet they clung to their ancient traditions of flint-working. While there is no clear understanding of the exact forces involved, farming spread rapidly across the face of Europe, probably reaching the coastal areas of southern Britain and Ireland before 4000 B.C.

Even by this date, hunting and gathering were by no means extinct

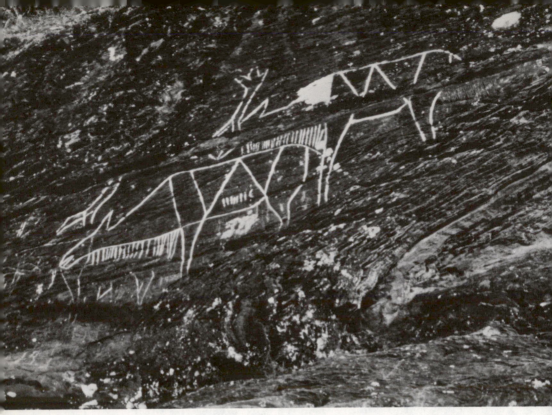

A pair of reindeer engraved in the Arctic rock art style near Stjördal in the
Trondelager valley, Norway.

in Europe: there are extensive regions, especially north of the
deciduous forest frontier, where fishing and sealing remain vital
economic activities to this day. Nor were the artistic impulses of the
hunter-fishers extinguished while great stone monuments, the megalithic tombs and temples, were being erected by the mass labor of
farming communities to the south. There was a spectacular revival
of animal art along the Arctic coastline of Norway, for example.
Many experts agree that these large rock engravings of reindeer, elk,
and hunting scenes have no connection with Ice Age art, but instead
belong to a tradition that began in about 3000 B.C. and flourished
in the north for many centuries.

It is also certain that isolated pockets of the old hunting cultures
persisted in areas that must have counted as unfavorable, marginal
land for the neighboring farmers. On the infertile sandy soils which
lie over wide areas of the Paris Basin, the rhythm of life of the
postglacial hunters continued, while all around them farming settlements became firmly established and expanded. Sandstone rock shelters were inhabited by these relic populations in places which are still
barren and desolate today, such as the Forest of Fontainebleau, and
the dates of occupation are as late as 3000 B.C. At this time, there
was a bizarre reawakening of the artistic spirit which had lain dor-

mant among most postglacial hunters for so long. On the walls and roofs of the sandstone shelters and caves, strange abstract patterns were carved in profusion, mainly wheel and star shapes, crosses and grills or grids, in a few cases covering the entire inside surfaces of the overhangs. There could be no greater contrast to the inspiration of Ice Age art than the stark vision of these last engravings of the hunters, with their rigid, regimented patterns.

This is where we must leave the record of hunting and gathering in prehistoric Europe, at an end in nearly all the temperate regions of the continent. Farming made possible achievements on a scale which no hunting economy, however complex, could possibly support—the erection of monuments such as Stonehenge and great palaces like Mycenae and Knossos. Farming also permitted the subjection of millions and the upward spiraling of the world's population, which some now estimate stands at nearly four hundred times that at the end of the Ice Age.

An example of the bizarre abstract rock art of the Paris Basin, probably dating to the Neolithic period. The site is called the Abri Leuillet.

It is relatively easy to suggest tht the invention of agriculture represents the true "Fall of Man," the fatal departure from the most successful adaptation in human history, an adaptation which for two million years or more had ensured a relatively secure and equitable existence for all. Inspired by Marshall Sahlins, we may mourn the loss of leisure, freedom, and prosperity enjoyed by the "Original Affluent Society."

On the other hand, to assault the values of our own society with those of a long-vanished past provides neither an effective criticism of those values nor a real understanding of what it was like to be a Paleolithic hunter. If there is any inherent point to the increasingly precise reconstructions of Ice Age landscapes provided by the pollen and bone specialists, or to the growing thoroughness with which the symbols of prehistoric art are recorded, it may be to bring home how strange and remote this world is to our imagination. As new scientific research shows how little the Paleolithic environment resembled any known countryside today, and how no single rationalization will satisfactorily explain the meaning of cave art, the experience of the Ice Age becomes not more but less accessible to us. The fascination of the Ice Age is that of the exotic and the unfamiliar; the challenge to our scientific methods and comprehension, or, if you prefer, pure escapism.

14: The Fading Image

As the guide and I climbed the path toward the rock shelter, we heard low scraping noises and the ring of metal against stone. At our approach, the sounds stopped. A hand-lamp flickered in the darkness ahead, its glow disappearing into one of the many passageways that entered the overhang from the outside. By the time we stepped into the shade of the rock, the intruder had gone, leaving behind a freshly disturbed heap of earth at our feet. Fortunately this assault had taken place in an area of the cave that had already been excavated, yet the clandestine digger had chosen no ordinary spot for his activities. We were standing under the low vault of La Vache in the Pyrenees, which is perched in the side of the Ariège valley. On a nearby slope, concealed behind stunted trees on the steep hillside, was the ancient entrance to Niaux, the immense system of underground tunnels which eventually leads after a distance of more than a kilometer to the most famous of all cave paintings outside of Lascaux and Altamira. It seems that the artists who painted the great black beasts of the Salon Noir never lived at Niaux, but they may well have occupied the shelter on the valley slope not far away. Here, at La Vache, the Magdalenians had discarded bone objects engraved with animal designs of exquisite quality almost matching the vigor and majesty of the Niaux paintings. These decorated objects had been patiently recovered during many years of controlled excavations at La Vache prior to the rummaging of the person whom we had surprised.

The mania for plundering archaeological sites is not new nor is it an isolated phenomenon, for it was the activity of enlightened collectors like Piette which helped to prepare the way for modern methods of scientific inquiry. But a century later, when the proper investigation of a prehistoric site demands months if not years of systematic teamwork, there can be no excuse for the indiscriminate actions of individuals.

Every cave which contains Paleolithic engravings or paintings is now protected by a metal grill or gate kept locked by a guide who is officially entrusted with the key. Despite such precautions, a few people are determined to enter at any cost. Just a day before the episode at La Vache and not too far away, the owner of one of the most frequently vandalized decorated caves had led us to its entrance, where we found that someone had recently tried to lever the gate off its hinges. We heard of other attempts which had been more

Graffiti on the metal door barring access to the decorated gallery at Les Eglises in the central Pyrenees.

successful. At Baume-Latrone in the Rhône Valley, the gate was blown off with dynamite, and the animal designs traced by a finger in wet clay deep inside the cave were deliberately destroyed.

This type of vandalism is not confined to well-known caves in obvious locations. The guide at an obscure site in northern Spain, located far from any main road, may well have exaggerated when he claimed to have replaced the padlock thirty times because of vandals. Nevertheless we found the gate swinging open, and after an exhausting crawl through cramped and partly flooded galleries, the visit proved disappointing. Of the eight animal designs outlined on a clay roof and copied by Breuil in 1906, only one had survived the daubings of later visitors. Even this solitary beast had its back line partially smudged by a modern finger. Comparable damage has been inflicted in other more richly decorated caves, notably at Gargas in the Pyrenees.

Stupidity and thoughtlessness play as big a part in this vandalism as does malicious intention. One morning I stood in the enormous echoing hall of Castillo, in northern Spain, at the rear of a coach party of about forty tourists. No guide, however conscientious or vigilant, could possibly supervise the actions of every person in such a group as they pass through the labyrinth of corridors and chambers at the site. The visitors had, however, been warned clearly not to touch any walls or ceilings. There we were, pausing underneath one of the largest painted compositions known in Ice Age art—the great frieze of animals, signs, spots, and hands executed in an unusual range of ocher shades. As the guide turned to lead the party to the next chamber, I watched in disbelief as first one hand and then another stretched up to finger the pigment on the ceiling. Would such people dream of reaching out to prod the canvas of an Old Master in an art museum?

Some degree of careless damage is probably an inevitable consequence of opening ancient monuments to the public. A more serious and specific problem is the changes which any human presence, no matter how cautious or infrequent, brings about in a cave. The sense of absolute immobility and timelessness which so often impresses a visitor is, of course, a total illusion. Every limestone cavern is alive with its own particular cycle of ceaseless corrosion and deposition; it "breathes" in a characteristic rhythm of atmospheric exchanges with the outside world. Moreover, the differences in humidity and temperature act in quite different ways from one area of the cave to the next, so that each passage or chamber has its own unique local "climate" and circumstances of rock growth or decay.

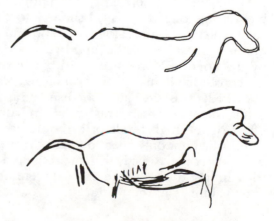

A clay drawing of a horse in a passage at Gargas as it existed in 1973, and *(below)* Breuil's record of the same design in 1911. Many other clay drawings in Gargas have deteriorated, chiefly because of damage by visitors.

Even a cave that is virtually sealed from the outside, such as was Lascaux before its discovery, is in no sense dead. At Lascaux, it has been shown, for example, that fluctuations recorded by thermometers placed inside the cave correspond to seasonal variations in temperature measured outside, except that there are differences of over one degree depending on exactly where the thermometers are planted within the layout of chambers and passageways. There is also a variation in temperature, depending on the height at which the inside measurements are registered; furthermore, in certain positions there is a time lag of several months between the inside and the outside climate. The important point is that Lascaux, which, at the time of its discovery, was hailed for its almost miraculous state of preservation, was not hermetically sealed like an airtight jar. Instead, like all caves, it was in a state of balance, an equilibrium of atmospheric forces. Indeed, there were large areas of Lascaux where paintings had entirely vanished from the walls through natural corrosion, just as in other areas the pigments shone with all their original freshness and brilliance.

Into this confined and delicately poised world, a human being is a drastic intrusion. He directly affects all the most important factors that control the atmosphere of the cave. The average visitor, for example, breathes out enough carbon dioxide during a single hour to produce, in an environment full of water, twenty liters of carbonic gas, the agent which attacks and dissolves limestone. In the same period of time, the visitor will generate sixty calories of body heat and forty grams of water vapor, and so the temperature and absolute humidity of the cave will rise. The impact of tourists on decorated caves which have been opened to the public is not, then, just a matter of casual physical damage. Indeed, the entire nature of the cave system can be altered in unforeseeable ways.

Even at Niaux, which is among the most spacious caverns at present accessible to tourists, the changes in the atmospheric balance of the cave due to its popularity during the summer must be considerable. During one such visit, I shone a torch upward at the immense vault which spans the Salon Noir, and realized that I could not make out its highest reaches because of a fog that hung there, generated by the continual presence of forty sweaty bodies far below. The effects of such human groups on more confined spaces can easily be imagined: the entire pattern of disintegration of wall surfaces and the formation of new ones is disrupted. At Lascaux, it was shown that the effect of the high levels of carbonic gas created by thousands of visitors had not only been to attack the walls of the cave but the gas had also reacted directly with the red pigments formed of iron oxide, thus dulling the brilliance of what were once the most vivid prehistoric paintings in the world.

This sculptured relief at La Chaire à Calvin in the Charente region of western France has been exposed to the open air for many years. Careful comparison of this recent photograph with earlier ones suggests that details of the relief have been lost through weathering.

Lascaux has also become notorious for another phenomenon, the so-called "Green Sickness" which invaded the decorated walls and ceilings. It is not only the atmosphere of caves which is constantly alive, for bacterial and plant life will flourish wherever light is present. In theory, the painted passages and chambers should remain in total darkness as much as possible, and the shoes of visitors should be disinfected to prevent the incursion of spores and microbes from the outside world. Certain organisms appear to be surprisingly responsive even to brief and localized exposures to light, so that the repeated brilliance of electronic flashes from tourist cameras may constitute a serious risk to the painted panels at which they are aimed. For this reason, flash photography is now banned in the painted caves, even for specialist researchers.

Can the combined effect of atmospheric and biological pollution be successfully overcome? This was the question which faced a commission of experts appointed in 1963 to save Lascaux, the "Sistine Chapel of Prehistory," from a condition of severe contamination. The melancholy history of tourist operations there has already been touched upon in reference to the concrete floors which were laid down directly over archaeological deposits without an attempt to investigate them. In some places the floor levels were also lowered and drained, affecting the humidity of the cave, while large bronze double doors were fitted, drastically altering the air circulation. After more than a decade of public viewing, the alarming increases in temperature and gas impurities led to the installation of air conditioning, which appeared to be a satisfactory measure for a while. In 1960, however, the growth of algae, the "Green Sickness," was first noted, and the inexorable spread of these organisms, together with the fading of ocher pigments on certain panels, finally led to the closing of Lascaux three years later. The seriousness of the contamination was revealed by the fact that even after light and visitors had been entirely excluded from the cave, both bacterial and plant life actually increased for a time. No alternative was left to the commission of experts but to disinfect the air and all exposed surfaces. The

Cave surfaces are in a constant state of change. This shows a wild ox and two ibexes in red and rows of spots at top painted in yellow not long after their discovery in March 1963 at La Tête du Lion, Bidon, in southeast France. The white patches at the top are calcite growths which obliterate cave paintings.

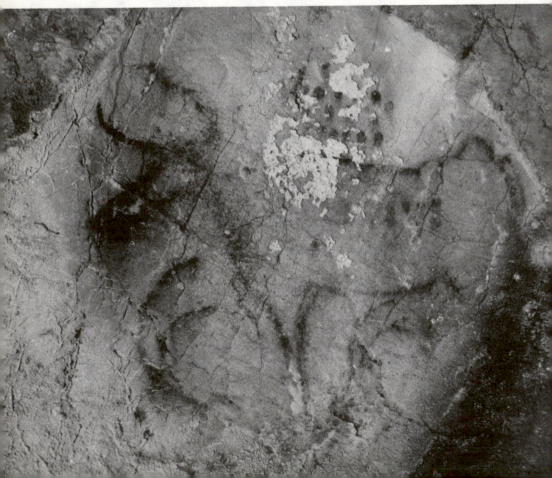

exposed surfaces were sprayed with a variety of substances carefully prepared to destroy the pollution but to leave the paintings unharmed. The circulation of atmospheric currents was controlled by the installation of airtight doors that divided off the three major sections of the cave. In addition, a machine was devised to create an artificial cold spot in the entrance hall.

After all these efforts, the slowly mounting damage which the presence of thousands of tourists had inflicted was finally brought to a halt. Today silence and darkness reign through the cave, broken only by a handful of strictly controlled visits, when subdued lamps are briefly switched on to illuminate the painted chambers. The exclusion of the casually interested tourist is the price that must be paid for the continuing preservation of Lascaux. No other remedy has as yet been devised which would not drastically alter the balance of temperature, humidity and atmosphere in the cave, nor expose the paintings to the risk of further pollution by microscopic agents.

The case of Lascaux has been discussed in detail not only because it reflects some of the problems shared by all decorated caves but also because it is one of the only sites where these problems have been studied at all. It is the world-wide fame of Lascaux and Altamira which, one hopes, has ensured their protection from further decline, whereas at dozens of other caves tourist exploitation continues unabated with the number of visitors rising dramatically each summer. In most cases, some token effort is made to restrict the number of visitors admitted at one time, but in reality the long-term effects of any human presence at each decorated cave are unknown and unpredictable. Already certain of the publicly accessible caves have unquestionably deteriorated over the years (there are some famous names amongst these), and it may be only a matter of time before all are doomed to corrosive or bacterial extinction.

What, then, is to be done? Is shutting down all the painted caves and never permitting ordinary persons to see cave art in its original context the right answer? The justice as well as the feasibility of such a ban is certainly debatable. At the moment the chaotic mixture of private exploitation and public powers of control in France and Spain ensures that the exceptionally famous or fragile caves are preserved, while others will be sacrificed to popular demand. In any case, the depth of the conservation problem means that a simple switch in ownership or access would not be the total answer. Probably the most rigorously protected of all the decorated sites have been the Volp caverns, which have remained in the hands of the Bégöuen family since their discovery just before the First World War. They have never been open to the public, and only a small number of privileged prehistorians has ever been conducted around the tortuous passages that lead to the clay bisons or the "Sorcerer." Despite

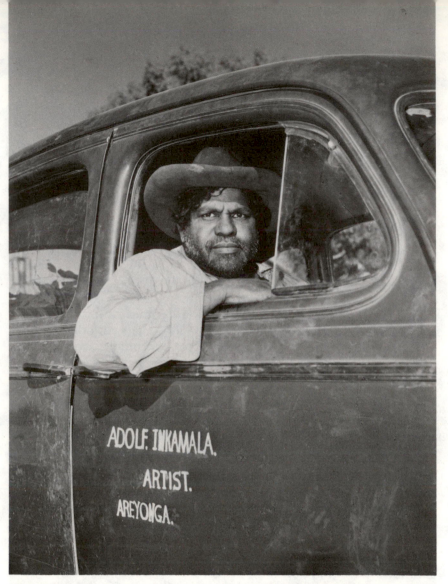

The tribal artist in the modern world. Adolf Inkamala, who works on a cattle ranch near Alice Springs, was one of the first Australian Aborigine artists to find acceptance on the European market with his realistic watercolors of bush scenes. He bought the car from the proceeds of his pictures.

this, it seems that the delicate engravings of the "Sanctuary" of Les Trois Frères have seriously declined since Breuil made his master copies many years ago. Even the painted outlines of the "Sorcerer" itself have deteriorated. Clearly, in certain cases any degree of human presence is incompatible with the continued existence of Ice Age art, and therefore some paintings and engravings have a limited lifespan from the moment of their discovery.

It is not the intention of this book to point a moral, since it, too, will be implicated in encouraging the growth of tourist pressure on the caves. Yet surely there exists an obvious responsibility to record properly what may soon be lost forever. The lack of serious academic research underground will probably come as a surprise to the general reader, who may suppose that these famous monuments have long ago been thoroughly planned and recorded for posterity. Though many inventories and monographs devoted to particular caves do exist, few of these (some would say less than a dozen) represent truly exhaustive efforts of description and analysis, while certainly the potential for the application of new methods of study (such as infrared and ultraviolet photography) is almost untouched.

It is extraordinary, for instance, that no chemical analysis has ever been published of any sample of pigment from a Paleolithic painting (apart from one case where tests were made to establish the authenticity of the art discovered at the Rouffignac cave in 1956). The destruction of a small area of some second-rate artwork might be weighed against the possibility of a positive gain in our knowledge of the procedures of cave artists, and perhaps also of the problems of conservation.

Indeed, the multidisciplinary work which has been an essential part of prehistoric archaeology for decades is a relative newcomer to work on the painted caves. As long ago as 1901, Professor Henri Moisson, the first chemist to receive the Nobel prize, pointed out, after the discovery of Font-de-Gaume, that natural destructive forces such as air currents, plant activity, light, and humidity, should be considered when examining the existing layout of artworks within a cave. Yet, during the past decade, only a handful of studies have appeared which include the reports of specialists in the geological sciences on the pattern of natural rock formation and decay unique to each cave. The co-operation of specialists outside the conventional realm of archaeology takes time and money, but such collaboration is commonplace in the study of excavated sites. In fact, it is the traditional glamour and prestige of excavation which undoubtedly attracts graduate students away from the exacting task of recording prehistoric art underground. As long as this is the case, systematic investigations of the world's oldest art will be few and far between.

So we are left with a concluding irony. The natural sciences have equipped the archaeologist with an unprecedented battery of new techniques, and the ability to ask new questions and answer old ones on a scale undreamed of by an older generation of scholars. At exactly this moment of unparalleled opportunity, the raw evidence itself is fast disappearing through inadequate protection and neglect. A century from now, will the present age of inquiry seem as barbarous as the days of Piette, whose workers threw painted pebbles in the river at Le Mas d'Azil for sport? This is the dilemma which has always faced archaeologists everywhere, but never more so than today to those who concern themselves with the secrets of the Ice Age.

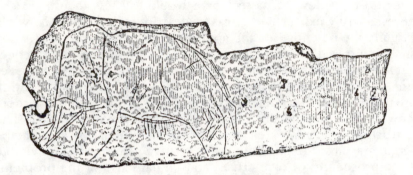

Mammoth engraving from Maltà, Siberia.

An Analysis of Animal Themes in Paleolithic Cave Art

by Antony Stevens

When this book was being set in type, I decided to attempt a new analysis of the animal themes in cave art. This is a simple type of study like that described in Chapter 10 (see pp. 211-213).

From published descriptions I obtained figures for how frequently ten different kinds of animal appear in each of forty caves (Fig. 1). Many of the things depicted in cave art, and many caves, had to be left out for various reasons (explained fully in my 1975 paper[1]). I wanted to see whether any of the caves had similar frequencies of animals.

"Cluster analysis" is a collection of numerical procedures suitable for this kind of problem. The first step is to find some mathematical formula that gives a high value when two caves are similar and a low value when they are not. The next step is to connect the caves in a tree diagram such that caves close together on the tree will be similar. We then look at the diagram (or "dendrogram," as it is called) to see if it suggests anything of interest.

There are many possible mathematical formulae to choose from, and I ran the computer program with nine different ones. Then I subjectively chose the one giving results that disagreed least with the others. The tree of connections is shown in Fig. 2.

We can now compare the new results with those of my earlier study. The connection of the three largest Spanish caves (Altamira, Castillo, Pasiega) is still close. The other interesting result of the previous test was that four large caves in the Pyrenees came out together (Marsoulas, Montespan, Le Portel, Niaux). This group has now "lost" Niaux. The reader is invited to study the data himself to form his own conclusions.

The tentative nature of this analysis cannot be emphasized too much. It is based on inventories of varying reliability and takes no account of such factors as the size, color(s), style or placing of the artworks within the cave.

I apologize to the reader for not solving any mystery.

[1]Stevens, A., 1975. *Association of Animals in Palaeolithic Cave Art: The Second Hypothesis,* in *Science and Archaeology* no. 16, p. 3.

Notes

The computer program used for this study is the widely available CLUSTAN (Version 1C, Release 2, D. Wishart, Edinburgh University 1978); the coefficient selected was 28. This work was partly supported by the Department of Mathematics and Statistics, City of London Polytechnic. A full list of references may be obtained by writing to the author at this Department.

	Number of:	HORSE	BISON	LARGE BOVID	SMALL BOVID	REINDEER	OTHER DEER	MAMMOTH	RHINO	BEAR	FELINE
Altamira		25	37	7	11	0	53	0	0	0	0
Altxerri		4	42	1	5	6	2	0	0	0	0
Bara-Bahau		8	2	3	0	0	0	0	0	1	1
Bédeilhac		6	25	0	0	1	0	0	0	0	0
Bernifal		4	2	0	0	0	0	11	0	0	0
Buxu		8	1	0	3	0	7	0	0	0	0
Candamo		11	5	15	6	0	10	0	0	0	0
Cap Blanc		6	3	0	0	0	0	0	0	0	0
Castillo		24	25	15	16	0	66	1	0	0	0
Chimeneas		2	0	10	5	0	13	0	0	0	0
Combarelles		116	37	7	9	14	9	13	1	19	5
Cougnac		0	0	0	7	0	5	4	0	0	0
Eglises		5	3	0	6	0	0	0	0	0	0
Ekain		34	11	0	5	0	3	0	0	2	0
Etcheberri		12	2	0	2	0	0	0	0	0	0
Font-de-Gaume		43	84	11	5	15	0	17	2	1	1
Gabillou		56	12	18	8	21	1	0	0	3	3
Gargas		43	35	19	15	0	9	6	0	0	0
Gazel		6	1	0	3	0	0	0	0	0	0
Hornos		12	11	5	4	0	1	0	0	0	0
Labastide		11	4	0	0	0	1	0	0	0	1
Marsoulas		32	32	1	4	0	1	0	0	0	0
Monedas		12	1	0	3	4	0	0	0	1	0
Montespan		27	23	0	0	0	0	0	0	1	4
La Mouthe		8	6	4	4	6	1	7	3	0	0
Niaux		30	55	3	16	0	3	0	1	0	0
Niño		2	0	0	3	0	5	0	0	0	0
Pair-non-Pair		14	0	12	8	2	3	5	1	5	1
Pasiega		45	20	13	14	0	57	1	0	0	0
Pech-Merle		6	10	10	0	0	4	13	0	1	1
Pergouset		3	3	0	3	0	4	0	0	0	0
Pindal		9	11	0	0	0	3	2	0	0	0
Portel		33	25	2	3	0	0	0	0	0	0
Rouffignac		13	21	0	14	0	0	117	11	2	0
Santimamiñe		6	26	1	2	0	1	0	0	1	0
Ste. Eulalie		5	0	0	3	4	1	0	0	1	0
Teyjat		9	3	3	0	18	10	0	0	2	0
Tito Bustillo		17	1	6	1	1	5	1	0	0	0
Trois Frères		66	178	11	23	23	12	4	1	10	9
Villars		7	2	0	1	0	0	0	0	0	0

Fig.1

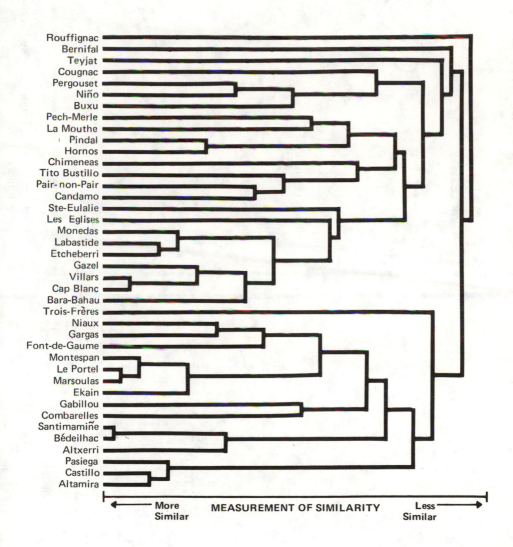

Fig.2

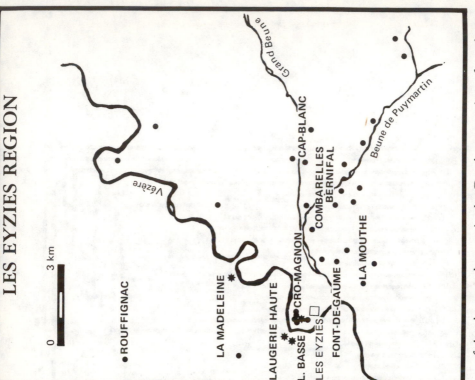

LES EYZIES REGION

0   3 km

ROUFFIGNAC

Vézère

Grand Beune

LA MADELEINE

LAUGERIE HAUTE

CAP-BLANC

L. BASSE   CRO-MAGNON

COMBARELLES

LES EYZIES   BERNIFAL

FONT-DE-GAUME

LA MOUTHE

Beune de Puymartin

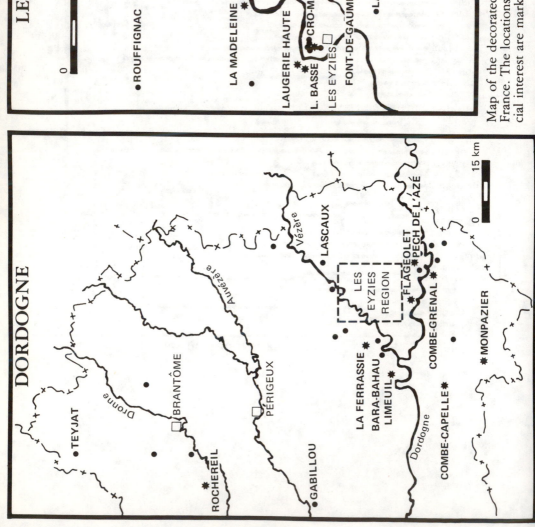

DORDOGNE

TEYJAT

BRANTÔME

Dronne

PÉRIGUEUX

Auvézère

Vézère

ROCHEREIL

GABILLOU

LA FERRASSIE

BARA-BAHAU

LIMEUIL

Dordogne

LASCAUX

LES EYZIES REGION

FLAGEOLET

PECH DE L'AZÉ

COMBE-GRENAL

COMBE-CAPELLE

MONPAZIER

0   15 km

Map of the decorated caves in the Dordogne region of southwest France. The locations of a few undecorated occupation sites of special interest are marked with stars.

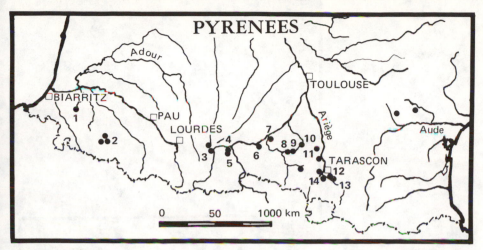

Map of the decorated caves in the French Pyrenees. 1. Isturitz, 2. Etcheberri, 3. Labastide, 4. Tibiran, 5. Gargas, 6. Montespan, 7. Marsoulas, 8 and 9. Les Trois Frères and Le Tuc d'Audobert, 10. Le Mas d'Azil, 11. Le Portel, 12. Les Eglises, 13. Fontanet, 14. Niaux.

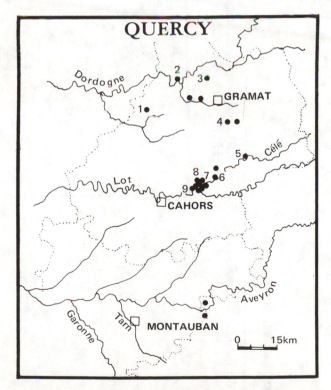

Map of the decorated caves in the Quercy region of France. 1. Cougnac, 2. Le Bourgnétou, 3. Les Fieux, 4. Les Escabasses, 5. Ste. Eulalie, 6. Le Papetier, 7. Cuzoul de Mélanie, 8. Pech-Merle, 9. Pergouset.

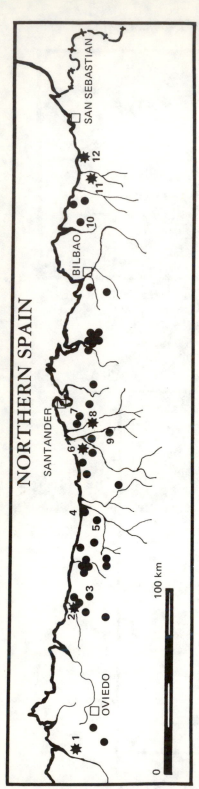

NORTHERN SPAIN

OVIEDO

SANTANDER

BILBAO

SAN SEBASTIAN

0   100 km

Map of the decorated caves in northern Spain. The most important sites are marked with a star. 1. La Peña de Candamo, 2. Tito Bustillo, 3. Buxu, 4. Pindal, 5. La Loja, 6. Altamira, 7. El Pendo, 8. El Castillo, La Pasiega, Las Chimeneas, Las Monedas, 9. Hornos de la Peña, 10. Santimamiñe, 11. Ekain, 12. Altxerri.

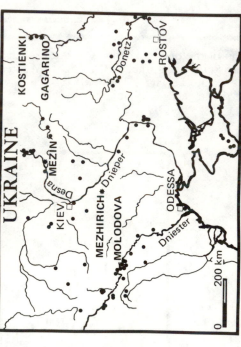

UKRAINE

KOSTIENKI

GAGARINO

MEZIN

Desna

KIEV

MEZHIRICH ∗ Dnieper

MOLODOVA

Dnieper

Donetz

ROSTOV

ODESSA

Dniester

0   200 km

The main Paleolithic occupation sites of all periods in the
Russian Ukraine. Those mentioned in the text are

Bibliography and Quotations

The following selection of works is not an attempt to provide an exhaustive reading list for the subjects covered in this book, but instead is a personal choice of references that the author has found particularly interesting or useful. The selection is divided into sub-headings related to the content of each chapter, with notes intended as further guides for the general reader. The more common periodicals are abbreviated as follows:

AmAnth	American Anthropologist
AmAntiq	American Antiquity
AmSci	American Scientist
Ann de Pal	Annales de Paléontologie
Ant	Antiquity
ArcAnth	Arctic Anthropology
BSPF	Bulletin de la Societé Préhistorique Française
CA	Current Anthropology
EtPréh	Etudes Préhistoriques
GalliaPréh	Gallia Préhistoire
HumEc	Human Ecology
JAnthRes	Journal of Anthropological Research
JArchSci	Journal of Archaeological Science
JHumEvo	Journal of Human Evolution
L'Anth	L'Anthropologie
PPS	Proceedings of the Prehistoric Society
PréhAr	Préhistoire Ariégoise
ProcRoyalSoc	Proceedings of the Royal Society
Quat	Quaternaria
Riv	Rivista di Scienze Preistoriche
SAArchBull	South African Archaeological Bulletin
ScAm	Scientific American
TrBrCaveResAssoc	Transactions of the British Cave Research Association
UISPP IX Cong	Union Internationale des Sciences Préhistoriques et Protohistoriques, Ninth Congress, Nice, 1976 (Prétirages)
WArch	World Archaeology

Chapter One

General works on prehistory and on hunter-gatherers

Perhaps the most readable introductions to prehistory are Pfeiffer 1972 and Clark and Piggott 1965. Two essential sources, on which this book draws heavily throughout, are Lee and DeVore 1968 and de Lumley 1976.

Bhattacharya, D. K., 1977. *Palaeolithic Europe,* Humanities Press, New Jersey.

Bicchieri, M. G., ed., 1972. *Hunters and Gatherers Today,* Holt, Rine-heart & Winston, New York.

Bourdier, F., 1967. *Préhistoire de France,* Paris.

Campbell, J. B., 1977. *The Upper Palaeolithic of Britain,* Oxford University Press, Oxford.

Clark, J. G. D., 1975. *World Prehistory in New Perspective,* Cambridge University Press, Cambridge.

    and Piggott, S., 1965. *Prehistoric Societies,* Penguin, London.

Coles, J. M., and Higgs, E. S., 1969. *The Archaeology of Early Man,* Penguin, London.

Fagan, B., ed., 1976. *Avenues to Antiquity, Readings from ScAm,* W. H. Freeman and Co., San Francisco.

Lee, R. B., and DeVore, I., eds., 1968. *Man the Hunter,* Aldine, Chicago.

Lumley, H. de, ed., 1976. *La Préhistoire Française,* CNRS, Paris.

Pfeiffer, J. E., 1972. *The Emergence of Man,* Harper and Row, New York (2nd edtn.).

Piggott, S., Daniel, G., and McBurney, C. B. M., eds., 1974. *France Before the Romans,* Thames and Hudson, London.

Pilbeam, D., 1972. *The Ascent of Man,* Macmillan, London.

Renfrew, C., ed., 1973. *The Explanation of Culture Change: Models in Prehistory,* Duckworth, London.

Waechter, J. D., 1976. *Man Before History,* Elsevier-Phaidon, Oxford.

General works on Paleolithic art

The best-illustrated book on cave art is Leroi-Gourhan 1968 (a new edition is imminent); the most important discussion is still Ucko and Rosenfeld 1967. For mobile art, see Marshack 1972.

Berenguer, M., 1973. *Prehistoric Man and His Art,* Souvenir Press, London.

Breuil, H., 1952. *Four Hundred Centuries of Cave Art,* Montignac.

    and Berger-Kirchner, L., 1960 (2nd edtn. 1970). *Franco-Cantabrian Rock Art,* in Bandi, H. G., ed., *The Art of the Stone Age,* London.

Cerda, F. J., Perello, E. R., et al., 1975. *La Prehistoria en La Cornisa Cantábrica,* Instit. de Preh. y Arq. Sautola, Santander.

Drouot, E., 1973. *Les Motivations de l'Art Paléolithique,* in *EtPréh,* Soc. Préh. de l'Ardèche, No. 4, p. 17.

Graziosi, P., 1960. *Palaeolithic Art,* Faber, London.
　　　　1973. *L'Arte Preistorica in Italia,* Sansoni, Florence.

Grigson, G., 1957. *The Painted Caves,* London.

Laming-Emperaire, A., 1962. *Lascaux,* Pelican, London.
　　　La Signification de l'Art Rupestre Paléolithique, Paris.

Leroi-Gourhan, A., 1968. *The Art of Prehistoric Man in Western Europe,* Thames and Hudson, London.

Marshack, A., 1972. *The Roots of Civilization,* McGraw, New York and Weidenfeld, London.

Santander Symposium 1972. *Actas del Symp. Int. de Arte Prehistorico,* Santander.

Saundars, N. K., 1968. *Prehistoric Art in Europe,* Penguin, London.

Sieveking, A., (in the press). *The Cave Artists,* Thames and Hudson, London.
　　　and G., 1962. *The Caves of France and Northern Spain,* London.

Ucko, P. J., and Rosenfeld, A., 1967. *Palaeolithic Cave Art,* World University Library, London.

Aspects of Paleolithic research: early work, east European evidence, population movements

Birdsell, J. B., 1977. *Populating Australia,* in Allen, J., Golson, J., and Jones, R., eds., *Sunda and Sahul,* Academic Press, London.

Bowler, J. M., Jones R., Allen, H., and Thorne, A. G., 1970. *Pleistocene Human Remains from Australia . . . ,* in *WArch* Vol. 2 No. 1, p. 39.

Brodrick, A. H., 1963. *The Abbé Breuil, Prehistorian,* London.

Clot, A., 1973. *L'Art Graphique Préhistorique des Hautes-Pyrénées,* Morlaas, St. Jammes. (see pp. 33–34 for account of Les Espélugues).

Klein, R. G., 1969. *Man and Culture in the Late Pleistocene: A Case Study,* W. H. Freeman and Co., San Francisco.
　　　　1973. *Ice Age Hunters of the Ukraine,* University of Chicago, Chicago.
　　　　1976. *Ice Age Hunters of the Ukraine,* in Fagan, B., ed., *Avenues to Antiquity, Readings from ScAm,* W. H. Freeman and Co., San Francisco.

Kozlowski, J. Z., and Kubiak, H., 1972. *Late Palaeolithic Dwellings Made of Mammoth Bones in South Poland,* in *Nature* Vol. 463 June 23rd, p. 237.

Lartet, E. and Christy, H., 1865–1875. *Reliquiae Aquitanicae,* London.

Leroi-Gourhan, A., and Brézillon, M., 1972. *Fouilles de Pincevent, VII supplément à Gallia Préhistoire,* CNRS, Paris.

MacNeish, R. S., ed., 1973. *Early Man in America,* W. H. Freeman and Co., San Francisco.

McBurney, C. B. M., 1976. *Early Man in the Soviet Union,* The British Academy, London.

Nelken-Terner, A., and MacNeish, R. S., 1977. *Séquences et Conséquences, ou des modalités américaines de l'adaptation de l'homme au Pleistocène,* in *BSPF* t. 74 Et. et Trav. fasc. 1, p. 293.

Orme, B., 1974. *Twentieth-Century Prehistorians and the Idea of Ethnographic Parallels,* in *Man* Vol. 9, p. 199.

Wendt, W. E., 1976. *Africa's Oldest Dated Works of Art,* in *SA Arch Bull* Vol. XXXI, p. 5.

Quotations

1. Leroi-Gourhan, A., 1968, p. 34.

Chapter Two

Ice Age geology and environments

Goudie 1977 is a comprehensive summary of recent research. Cornwall 1970 and Matsch 1976 are excellent popular accounts covering the same ground.

Bowen, D. Q., 1978. *Quaternary Geology . . . ,* Oxford and New York.

Butzer, K. W., 1972. *Environment and Archaeology,* Methuen, London, (2nd edtn.).

Cornwall, I., 1970. *Ice Ages: Their Nature and Effects,* Humanities, New Jersey.

Flint, R. F., 1971. *Glacial and Quaternary Geology,* Wiley, New York.

Goudie, A. S., 1977. *Environmental Change,* Clarendon Press, Oxford.

Matsch, C. L., 1976. *North America and the Great Ice Age,* McGraw, New York.

West, R. G., 1977, *Pleistocene Geology and Biology,* Longman, London, (2nd edtn.).

Human evolution and Lower Paleolithic archaeology

The books by Leakey and Lewin form attractive introductions for the general reader, while the article by Isaac is particularly recommended.

Bordes, F., 1969. *Os Percé Moustérien et Os Gravé Acheuléen . . . ,* in *Quat* Vol. XI, p. 1.

Butzer, K. W., and Isaac, G. Ll., 1975. *After the Australopithecenes,* Mouton, The Hague.

Clark, J. D., 1975. *Africa in Prehistory: Peripheral or Paramount?,* in *Man* Vol. 10, p. 175.

Edwards, S. W., 1978. *Non-Utilitarian Activities in the Lower Palaeolithic,* in *CA* Vol. 19 No. 1, p. 135.

Isaac, G. Ll., 1978. *The Food-Sharing Behavior of Protohuman Hominids,* in *ScAm* Vol. 238 No. 4, p. 90.

    and McCown, E. R., 1976. *Human Origins: Louis Leakey and the East African Evidence,* Benjamin, California.

Jolly, C., ed., 1978. *Early Hominids of Africa,* Duckworth, London.

Lancaster, Jane B., 1975. *Primate Behaviour and the Emergence of Human Culture,* Macmillan, New York.

Leakey, R. E. F., and Lewin, R., 1977. *Origins,* Macdonald, London, and Dutton, New York.

    1978. *People of the Lake,* Doubleday, New York.

Lumley, H. de, 1976. *A Palaeolithic Camp at Nice,* in Fagan, B., ed., *Avenues to Antiquity, Readings from ScAm,* W. H. Freeman and Co., San Francisco, p. 36.

    ed., *La Préhistoire Française,* CNRS, Paris. See pp. 548 and 631 for information on Arago finds.

Marshack, A., 1977. *The Meander as a System . . . ,* in Ucko, P. J., ed., *Form in Indigenous Art,* Australian Inst. of Abo. St., London and New Jersey, see pp. 291–292 for Pech de l'Azé bone.

Quotations

1. Agassiz, L., in Carozzi, A. V., ed., 1967. *Studies on Glaciers,* Hafner, New York.
2. Leakey, M. D., 1971. *Olduvai Gorge, Vol. 3, Excavations in Bed I and II, 1960–63,* Cambridge, p. 269.
3. Pilbeam, D. B., in Butzer, K. W., and Isaac, G. Ll., eds., 1975, p. 812.

Chapter Three

Neanderthal man

Although it is not widely known, the book by Kennedy is a concise, accurate and well-written general introduction to this topic. For the history of early research on the Neanderthals, see Daniel and Campbell. Two papers of particular importance for the current state of inquiry are Howells 1974 and Stringer 1974.

Bordes, F., ed., 1969. *The Origin of Homo Sapiens,* Unesco. See especially pp. 49 (Qafza burials) and 129 (Löwenburg flint mine).

Boule, M., and Vallois, H. V., 1957. *Fossil Men,* Thames and Hudson, London (one of several English editions).

Brace, C. L., 1964. *The Fate of the Classic Neanderthals,* in *CA* Vol. 5, p. 3.

Campbell, B. G., 1956. *The Centenary of Neanderthal Man,* in *Man* Vol. LVI, pp. 174 and 195.

Daniel, G. E., 1964. *The Idea of Prehistory,* London.
    1975. *150 Years of Archaeology,* London. (2nd edtn.).

Howells, W. W., 1974. *Neanderthals: Names, Hypotheses and Scientific Methods,* in *AmAnth* Vol. 76, p. 24.
    1975. *Neanderthal Man: Facts and Figures,* in Tuttle, R. H., ed., *Paleoanthropology; Morphology and Paleoecology,* Mouton, The Hague.

Kennedy, A. R., 1975. *Neanderthal Man,* Burgess, Minneapolis.

Koenigswald, G. H. R., ed., 1958. *Hundert Jahre Neanderthaler,* Utrecht.

Marshack, A., 1976. *Some Implications of the Paleolithic Symbolic Evidence for the Origin of Language,* in *CA* Vol. 17 No. 2, p. 274, and in *AmSci* Vol. 64 No. 2, p. 136.

Piperno, M., 1976–7. *Analyse du Sol Moustérien de la Grotte Guttari au Monte Circeo,* in *Quat* Vol. XIX, p. 71.

Solecki, R. S., 1971. *Shanidar,* Allen Lane, London.

Stringer, C. B., 1974. *Population Relationships of Later Pleistocene Hominids,* in *JArchSci* Vol. 1, p. 317.

Mousterian cultures

There is no popularly written and balanced account of the Mousterian debate available. To form an impression of it, see the papers referred to below in Renfrew 1973 and in *WArch* Vol. 2, 1970. For population movements in Europe, see Gábori 1976.

Binford, L. R., 1972. *An Archaeological Perspective,* Academic Press, New York and London.
    1973. *Inter-assemblage Variability,* in Renfrew, C., ed., *The Explanation of Culture Change,* Duckworth, London, p. 227.
    and S. R., 1966. *A Preliminary Analysis of Functional Variability . . . ,* in *AmAnth* Vol. 68 (2) Part 2, p. 238.
    eds., 1968. *New Perspectives in Archaeology,* Aldine, Chicago.
    1969. *Stone Tools and Human Behavior,* in *ScAm* Vol. 220, p. 70.

Bordes, F., 1968. *The Old Stone Age,* World University Library, London.
    1972. *A Tale of Two Caves,* Harper and Row, New York.
    1973. *On the Chronology and Contemporaneity of Different Pala-*

eolithic Cultures in France, in Renfrew, C., ed., *The Explanation of Culture Change,* p. 217.

Fine, D., 1977. *The Semantics of Mousterian Variability,* unpub. MSS.

Gábori, M., 1976. *Les Civilisations du Paléolithique Moyen Entre Les Alpes et L'Oural,* Akadémiai Kiadó, Budapest.

Lumley, H. de, 1972. *La Grotte Moustérienne de l'Hortus,* Univ. de Provence, Marseille.

Mellars, P. A., 1969. *The Chronology of Mousterian Industries . . . ,* in *PPS* Vol. 35, p. 134.

    1970. *Some Comments on the Notion of Functional Variability . . . ,* in *WArch* Vol. 2, p. 74.

Rolland, N., 1977. *New Aspects of Middle Palaeolithic Variability in Western Europe,* in *Nature* Vol. 266 March 17th, p. 251.

Quotations

1. Schaafhausen, H., 1857. *Verh. Naturh. Ver. Preuss. Rhenil.,* Bonn, Vol. XIV, pp. XXXVIII–XLII.
2. Lightfoot, c.f. footnote in Daniel, G. E., 1975, p. 27.
3. Bouyssonie, A. and J., and Bardon, L., as quoted by Vandermeersch, B., in Lumley, H. de, ed., 1976. *La Préhistoire Française,* CNRS, Paris, p. 725.
4. Straus, W. L., and Cave, A. J. E., 1957. *Pathology and the Posture of Neanderthal Man,* in *Quarterly Review of Biology,* December, p. 348.
5. Gerasimov, M., 1964. *The Face Finder,* London.
6. Binford, L. R., 1972, p. 192.
7. Bordes, F., 1968, pp. 144–145.
8. Marshack, A., 1976, p. 277.

Chapter Four

Aspects of Upper Paleolithic archaeology

For interesting accounts of the cultural sequences in France and Russia, see Piggott, Daniel and McBurney 1974, and McBurney 1976 respectively.

Arambourou, R., 1962. *Sculptures Magdaléniennes Découvertes à la Grotte Duruthy,* in *L'Anth* t. 66, p. 457.

    1973. *Les Gisements Préhistoriques de Sorde . . . ,* in *Préh. et Protoh. des Pyr.,* Lourdes.

    1977. *La Fin des Temps Glaciaires à Duruthy,* in *La Fin des Temps Glaciaires en Europe,* Colloq. Int. du CNRS No. 271, Instit. de Quat., Bordeaux, p. 24.

    (in the press). Memoir on Duruthy for the Soc. Préh. Fr.

Bader, O., 1970. *The Boys of Sungir,* in *Illustrated London News,* March 7th, p. 24.

Banesz, L., 1976. *Les Structures d'Habitat au Pal. Sup. en Europe Centrale,* in Colloque XIII *UISPP IX Cong.,* p. 8.

Drouot, E., 1968. *L'Art Paléolithique du Languedoc Mediterranéen,* in *La Préhistoire: Problèmes et Tendances,* CNRS, Paris, p. 145.

Lynch, T. F., 1966. *The 'Lower Perigordian' in French Archaeology,* in *PPS* Vol. 7, p. 156.

McBurney, C. B. M., 1976. *Early Man in the Soviet Union,* British Academy, London.

Megaw, J. V. S., 1960. *Penny Whistles and Prehistory,* in *Ant* XXXIV, p. 6.

Mellars, P. A., 1973. *The Character of the Middle-Upper Pal. Transition In South-West France,* in Renfrew, C., ed., *The Explanation of Culture Change,* Duckworth, London, p. 255.

Movius, H. J., 1969. *The Châtelperronian in French Archaeology,* in *Ant* Vol. XLIII, p. 111. (Account of Arcy-sur-Cure finds).

    1975. *Excavations of the Abri Pataud,* Peabody Museum, Cambridge, Mass. (First of a series of volumes).

Péquart, M. et S., 1960. Mas d'Azil, in *Ann de Pal* t. XLVI, see p. 243 for "tattooing" materials.

Piggott, S., Daniel, G., and McBurney, C. B. M., eds., 1974. *France Before the Romans,* Thames and Hudson, London.

Rigaud, J. Ph., 1976. *Données Nouvelles sur le Périgordien Supérieur en Périgord,* in Colloque XV *UISPP IX Cong.,* p. 53.

Sackett, J. R., 1968. *Method and Theory in Upper Palaeolithic Archaeology in S. W. France,* in Binford, L. R. and S. R., eds., *New Perspectives in Archaeology,* Aldine, Chicago, p. 61.

Smith, P., 1976. *The Solutrian Cultures,* in Fagan, B., ed., *Avenues to Antiquity, Readings from ScAm,* W. H. Freeman and Co., San Francisco.

Quotations

1. Henderson, K., 1927. *Prehistoric Man,* London, pp. 59 and 65.
2. Boule, M., and Vallois, H. V., 1957. *Fossil Men,* Thames and Hudson, London, p. 257.
3. Macalister, R. A. S., 1921. *A Textbook of European Archaeology,* Cambridge University Press, Cambridge, p. 348.
4. Breuil, H., 1935. Review of Peyrony in *L'Anth* t. 45, p. 115.
5. Arambourou, R., 1962, p. 467.

Chapter Five

Paleolithic economies

Both the Higgs volumes are the basis of much recent discussion and research in economic prehistory. Laville and Renault Miskovsky 1977 contains informative and entertaining discussions.

Bordes, F., *et al.,* 1972, on Würm II environments in Lumley, H. de, *La Grotte Moustérienne de l'Hortus,* Univ. de Provence, Marseille, p. 353.

David, N. C., 1973. *On Upper Palaeolithic society, ecology and technological change,* in Renfrew, C., ed., *The Explanation of Culture Change,* Duckworth, London, p. 277.

Delpech, F., and Rigaud, J. P., 1974. *Etude de la Fragmentation et de la Repartition des Restes Osseux,* in Camps-Fabrer, H., ed., *L'Industrie de l'Os dans la Préhistoire,* Univ. de Provence, Marseille, p. 47.

Freeman, L. G., 1973. *The Significance of Mammalian Faunas from Palaeolithic Occupations in Cantabrian Spain,* in *AmAntiq* Vol. 38, p. 3. (However, see Straus 1977 for up-dating of evidence).

Gábori-Csánk, V., 1968. *La Station du Paléolithique Moyen d'Érd, Hongrie,* Akadémiai Kiadó, Budapest.

Higgs, E. S., ed., 1972. *Papers in Economic Prehistory,* Cambridge University Press, Cambridge.

    1975. *Palaeoeconomy,* Cambridge University Press, Cambridge.

Jochim, M., 1976. *Hunter-gatherer Subsistence and Settlement: A Predictive Model,* Academic Press, New York and London.

Laville, H., and Renault Miskovsky, J., 1977. *Approche Ecologique de l'Homme Fossile,* Univ. Pierre et M. Curie.

Puech, P. F., 1976. *Recherche sur le mode d'alimentation . . . ,* in Lumley, H. de, ed., *La Préhistoire Française,* CNRS, Paris, p. 708.

Straus, L. G., 1976–7. *The Upper Palaeolithic Cave Site of Altamira, Spain,* in *Quat* Vol. XIX, p. 135.

    1977. *Of Deerslayers and Mountain Men: Faunal Exploitation in Cantabrian Spain,* in Binford, L. R., ed., *For Theory Building in Archeology,* Academic Press, New York and London.

Uerpmann, H. P., 1972. *Animal Bone Finds and Economic Archaeology,* in *WArch* Vol. 4, p. 307.

Woodburn, J., 1968. *An Introduction to Hadza Ecology,* in Lee, R. B., and DeVore, I., eds., *Man the Hunter,* Aldine, Chicago, p. 49.

Quotations

1. Czaplicka, M. A., 1914. *Aboriginal Siberia,* Oxford, p. 11.
2. Leroi-Gourhan, Arl., 1963. *Archéologie et Botanique,* as quoted in Lumley, H. de, ed., 1976. *La Préhistoire Française,* CNRS, Paris, p. 685.
3. Leeds, A., 1965. *Reindeer Herding and Chuckchee Social Institutions,* in Leeds, A., and Vayda, A. P., eds., *Man, Culture and Animals,* American Assoc. for Adv. of Science No. 78, Washington, p. 92.

Chapter Six

Horse and reindeer exploitation

As this book was going to press, two important publications appeared: an excellent history of the domestication debate, Bahn 1978, and the exciting discovery of a dog in Azilian layers dating to about 9600 B.C., Célèrier and Delpech 1978. For conflicting views on reindeer in prehistory, see papers by Burch and Sturdy.

Bahn, P. G., 1976. *Les Bâtons Percés . . . Réveil d'une Hypothèse Abandonnée,* in *PréhAr* t. XXXI, p. 47.
    1977. *Seasonal Migration in South-west France During the Late Glacial Period,* in *JArch Sci* Vol. 4, p. 245.
    1978. *The 'Unacceptable Face' of the West European Upper Palaeolithic,* in *Ant* Vol. LII, p. 183.
Binford, L. R., 1973. *Inter-assemblage Variability . . . ,* in Renfrew, C., ed., *The Explanation of Culture Change,* Duckworth, London, see pp. 238–240.
Bogoras, W., 1904. *The Chuckchee,* Jessup N. Pacific Expedition Mem. VII, The American Museum of Natural History.
Bouchud, J., 1954. *Le Renne et le Problème des Migrations,* in *L'Anth* Vol. 58, p. 79.
Burch, E. S., 1972. *The Caribou/Wild Reindeer as a Human Resource,* in *AmAntiq* Vol. 37 No. 3, p. 339.
Célèrier, G., and Delpech, F., 1978. *Un Chien dans l'Azilien . . . ,* in *BSPF* t.75, p. 212.
Chard, C. S., 1963. *The Nganasan—Wild Hunters of the Reindeer,* in *ArcAnth* Vol. 1, p. 105.
Ingold, T., 1974. *On Reindeer and Men,* in *Man* Vol. 9 No. 4, p. 523.
    1976. *The Skolt Lapps Today,* Cambridge University Press, Cambridge.
Leeds, A., and Vayda, A. P., 1965. *Man, Culture and Animals,* Am. Assoc. Adv. Science No. 78, Washington. See Leeds, A., on the Chukchi, p. 87, and Meggitt, M. J., on dingoes, p. 7.
Omnes, J., 1977. *Nouvelle Gravure de Tête d'Equidé dans la Grotte d'Espalungue,* in *BSPF* t. 74 CRSM No. 3, p. 89.
Pales, L., and St. Péreuse, M. de, 1966. *Un Cheval Prétexte—Retou du Chevêtre,* in *Objets et Mondes* t. 6, p. 187.
Piette, E., 1889. *L'Art Pendant l'Age du Renne—La Question de la Domestication du Renne,* in Cong. Int. d'Anth. et d'Arch. Préh., 10th Session, p. 159.
Sturdy, D. A., 1972. *The Exploitation Patterns of a Modern Reindeer Economy in Western Greenland,* in Higgs, E. S., ed., *Papers in Eco-*

nomic Prehistory, Cambridge University Press, Cambridge.

    1975. *Some Reindeer Economies in Prehistoric Europe,* in Higgs, E. S., ed., *Palaeoeconomy,* Cambridge University Press, Cambridge.

Quotations

1. Piette, E., 1889, p. 159.
2. Mowat, F., 1952. *The People of the Deer,* London, pp. 9–10 and 65.
3. Murie, O. J., 1935. *Alaska-Yukon Caribou, U.S. Dept. Agric. Bureau Biological Survey, N. American Fauna No. 54,* Washington, p. 43.
4. Shirokogoroff, S. M., 1933. *Social Organization of the Northern Tungus,* Shanghai, p. 30.
5. Sturdy, D. A., 1972, p. 167.

Chapter Seven

Attitudes to hunter-gatherers; the control of population

Note particularly the works by Sahlins and Lee and DeVore. For the general background to problems of population history, see Short 1976 and Wilkinson 1973.

Ammerman, A. J., 1975. *Late Pleistocene Population Dynamics: An Alternative View,* in *HumEc* Vol. 3 No. 4, p. 219.

Denham, W. W., 1974. *Population Structure, Infant Transport and Infanticide . . . ,* in *JAnthRes* Vol. 30, p. 191.

Divale, W. T., 1972. *Systemic Population Control in the Middle and Upper Palaeolithic,* in *WArch* Vol. 4, p. 222.

Hayden, B., 1972. *Population Control among Hunter-gatherers,* in *WArch* Vol. 4, p. 205.

Howell, N., 1976. *The Population of the Dobe Area !Kung,* in Lee, R. B., and DeVore, I., eds., *Kalahari Hunters and Gatherers,* Harvard University Press, p. 137.

    Towards a Uniformitarian Theory of Human Paleodemography, in *JHumEvo* Vol. 5 No. 1., p. 25.

Lee, R. B., and DeVore, I., eds., 1968. *Man the Hunter,* Aldine, Chicago. See articles by Lee, p. 30, Woodburn, p. 103, Birdsell, p. 229, and others.

    eds., 1976. *Kalahari Hunters and Gatherers,* Harvard University Press.

Sahlins, M., 1974. *Stone Age Economics,* Tavistock, London.

Schrire, C., and Steiger, W. L., 1974. *A Matter of Life and Death: An Investigation into the Practice of Female Infanticide in the Arctic,* in *Man* Vol. 9 No. 2, p. 161.

Short, R. V., 1976. *The Evolution of Human Reproduction,* in *ProcRoyal Soc,* London, Vol. B 195, p. 3.

Turgot, 1751. *On Universal History,* tr. Meek, L. R., 1973, Cambridge University Press, Cambridge.

Wilkinson, R. G., 1973. *Poverty and Progress,* London.

Williams, B. J., 1977. *A Question about !Kung Ecology,* in *Science* Vol. 198 No. 4319, 25th Nov, p. 782.

Yellen, J., and Harpending, H., 1972. *Hunter-gatherer Populations and Archaeological Inference,* in *WArch* Vol. 4 No. 2, p. 244.

Disease in prehistory

For a general introduction, see Brothwell 1972. Janssens 1970 is an interesting popular book, although somewhat unreliable.

Barrière, C., 1975. *L'Art Pariétal de la Grotte de Gargas, British Archaeological Reports,* Supp. Series No. 14. (With translation).

Brothwell, D. R., 1972. *Digging Up Bones,* British Museum. (2nd edtn.).

Clot, A., 1973. *L'Art Graphique Préhistorique des Hautes-Pyrénées,* Morlaas, St. Jammes, see pp. 102–109.

Dastuge, J., and Lumley, M.-A. de, 1976. *Les Maladies des Hommes Préhistoriques . . . ,* in Lumley, H. de, *La Préhistoire Française,* CNRS, Paris, p. 612.

Janssens, P. A., 1970. *Palaeopathology,* Humanities, New Jersey.

Pradel, L., 1975. *Les Mains Incomplètes de Gargas, Tibiran et Maltravieso,* in *Quartar* Bd. 26, p. 159.

Vallois, H. V., 1971. *Le Crâne Trépané Magdalénien de Rochereil,* in *BSPF* t. 68, p. 485.

Quotations

1. Turgot, 1751, p. 65.
2. Braidwood, R. J., and Howe, B., 1960. *Prehistoric Investigations in Iraqui Kurdistan, Univ. of Chicago Oriental Inst., Studies in Ancient Oriental Civilization,* p. 1.
3. Sahlins, M., 1974, p. 4.
4. Ibid., p. 36.
5. Ibid., p. 38.
6. Ibid., p. 23.
7. Schrire, C., and Steiger, W. L., 1974, p. 178.
8. Woodburn, J., 1968. As quoted in *Man the Hunter,* Lee R. B., and DeVore, I., eds., Aldine, Chicago, p. 91.
9. Hayden, B., 1972, p. 206.
10. Rasmussen, K., 1931. *The Netsilik Eskimos: Social Life and Spiritual Culture, Report of the 5th Thule Expedition 1921–1934,* Copenhagen.

Chapter Eight

The social organization of hunter-gatherers

The basic arguments for this chapter were suggested by two important papers, Bahn 1977 and Yellen in *WArch* 1977. Peterson's volume conveys a fascinating picture of the problems of Australian Aboriginal tribal research. Another key review of the Magdalenian evidence is Sieveking 1976, although it contains many factual errors.

This chapter touches superficially on some difficult anthropological problems. Among interesting new work too complex to be included here, see papers by Wobst.

Bahn, P. G., 1977. *Seasonal Migration in South-west France During the Late Glacial Period,* in *JArchSci* Vol. 4, p. 245.

Balicki, A., 1968. *The Netsilik Eskimos: Adaptive Processes,* in Lee, R. B., and DeVore, I., eds., *Man the Hunter,* Aldine, Chicago, p. 78.

Chollot, M., 1964. *Collection Piette,* Mus. des Ant. Nat., Paris.

Grigor'ev, G. P., 1967. *A New Reconstruction of the Above-Ground Dwelling of Kostenki I,* in *CA* Vol. 8 No. 4, p. 344.

Klein, R. G., 1973. *Ice Age Hunters of the Ukraine,* University of Chicago, Chicago.

Lee, R. B., 1972. *Work Effort, Group Structure and Land-use in Contemporary Hunter-gatherers,* in Ucko, P. J., Tringham, R., and Dimbleby, G. W., eds., *Man, Settlement and Urbanism,* Duckworth London, p. 177.

    1976. *!Kung Spatial Organization,* in Lee, R. B., and DeVore, I., eds., *Kalahari Hunters and Gatherers,* Harvard University Press, p. 73.

Oakley, K., 1965. *Folklore of Fossils,* in *Ant* Vol. 39, pp. 9 and 117.

Peterson, N., ed., 1976. *Tribes and Boundaries in Australia,* Instit. Abo. Studies, Canberra. See papers by Mulvaney, D. J., Birdsell, J. B., and Yengoyan, A. A.

Sieveking, A., 1976. *Settlement Patterns of the later Magdalenian in the Central Pyrenees,* in Sieveking, G., Longworth, I. H., and Wilson, K. E., eds., *Problems in Economic and Social Archaeology,* Duckworth, London, p. 583.

Spencer, B., and Gillen, F. J., 1899. *The Native Tribes of Central Australia,* London.

Vita-Finzi, C., 1974. *Age of Valley Deposits in Périgord,* in *Nature* Vol. 250, p. 568.

Wobst, H., 1974. *Boundary Conditions for Paleolithic Social Systems* . . . in *AmAntiq* Vol. 39 No. 2, p. 147.

    1976. *Locational Relationships in Paleolithic Societies,* in *JHumEvo* Vol. 5, p. 49.

Yellen, J. E., 1977. *Archaeological Approaches to the Present,* Academic Press, New York and London.

    Long-term Hunter-gatherer Adaptations to Desert Environments . . ., in *WArch* Vol. 8 No. 3, p. 262.

    and Harpending, H., 1972. *Hunter-gatherer Populations and Archaeological Inference,* in *WArch* Vol. 4 No. 2, p. 244.

Quotations

1. Bordes, F., 1968. *The Old Stone Age,* World University Library, London, p. 235.
2. Radcliffe-Brown, A. R., 1931. *Social Organization of Australian Tribes,* in *Oceania* Vol. 1, p. 35.
3. Yellen, J. E., and Harpending, H., 1972, as summarized by Yellen, 1977 in *WArch,* p. 267.
4. Lee, R. B., 1972, p. 182.
5. Balicki, A., 1968. *A Comparative Study on Subsistence Ecology and Social Systems . . . ,* in *Proc. of the VIII Cong. of Anth. and Ethn. Sciences,* Tokyo, p. 262.
6. Spencer, B., and Gillen, F. J., 1899, p. 587.
7. Berndt, R. M., 1972. *The Walmadjeri and Gugadja,* in Bicchieri, M. G., ed., *Hunters and Gatherers Today,* Holt, Rinehart & Winston, New York, p. 188.

Chapter Nine

Activities inside decorated caves

See Breuil and Bégouën 1958 for its marvelous illustrations.

Bégouën, M., 1925. *Les Bisons d'Argile,* Paris.
Bégouën, R., Clottes, J., and Delporte, H., 1977. *Le Retour du Petit Bison au Tuc d'Audobert,* in *BSPF* t. 74 CRSM No. 4, p. 112.
Berenguer, M., 1973. *Prehistoric Man and his Art,* Souvenir Press, London, see p. 164.
Breuil, H., and Bégouën, H., 1958. *Les Cavernes du Volp,* Trav. de l'Instit. de Pal. Hum., Paris.
Clottes, J., and Simmonet, R., 1972. *Le Réseau René Clastres de la Caverne de Niaux (Ariège),* in *BSPF* t. 69 Et. et Trav., p. 293.
Creer, K. M., and Kopper, J. S., 1974. *Paleomagnetic Dating of Cave Paintings in Tito Bustillo . . . ,* in *Science* Vol. 186, p. 348.
Delteil, J., Durbas, P., and Wahl, L., 1972. *Présentation de la Galerie Ornée de Fontanet,* in *PréhAr* t. XXVII, p. 11.
Glory, A., 1965. *Nouvelles Découvertes de Dessins Rupestres sur le Causse de Gramat,* in *BSPF* t. 62, p. 528.
Hooper, A., 1978. *Palaeolithic Cave Art and the Natural Lighting of*

Caves, in *TrBrCaveResAssoc* Vol. 5 No. 1, p. 13.
    and Collison, D., 1976. *L'Art Paléolithique de la Grotte des Eglises,* in *GalliaPréh* t. 19, p. 221.
Laming, A., 1962. *Lascaux,* Pelican, London.
Lorblanchet, M., 1971. *Nouvelles Figures Pariétales Paléolithiques en Quercy,* in *BSPF* t. 68 Et. et Trav., p. 293.
    1974. *L'Art Préhistorique en Quercy,* Morlaas, St. Jammes.

Quotations

1. Hooper, A., 1977. Interview with author.
2. Ibid.
3. Kuhn, H., 1955. *On the Track of Prehistoric Man,* London, pp. 84–85.
4. Bégouën, H., 1939. *Les Bases Magiques de l'Art Préhistorique,* in *Scientia,* p. 215.
5. Bégouën, M., 1925.

Chapter Ten

The meaning of Paleolithic cave art

The basic work is Ucko and Rosenfeld 1967. For recent reports of exceptional quality on decorated caves, see works by Altuna and Apellaniz, and Lorblanchet.

Altuna, J., and Apellaniz, J. M., 1976. *Las Figuras Rupestres Paleolíticas de la Cueva de Altxerri,* in *Munibe* Año XXVIII 1–3.
    1978. *Las Figuras Rupestres de la Cueva de Ekain,* in *Munibe* Año XXX 1–3.
Barandarian, I., 1975. *El Arte Mobiliar Cantábrico,* in *La Prehistoria en la Cornisa Cantábrica,* Instit. Cult. de Cantabria, Santander.
Hooper, A., 1976. *The Meaning of Palaeolithic Cave Art,* Unpublished Thesis, University of London.
    1977. *Note sur les Oiseaux Figurés dans l'Art Paléolithique,* in *PréhAr* t. XXXII, p. 85.
Laming-Emperaire, A., 1971. *Une Hypothèse de Travail pour une Nouvelle Approche des Sociétés Préhistoriques,* in *Mélanges Offerts à A. Varagnac,* Ec. Pratique des H. Etudes, Paris.
Leason, P. A., 1939. *A New View of the Western European Group of Quaternary Cave Art,* in *PPS* Vol. 5, p. 51.
Leroi-Gourhan, A., 1968. *The Art of Prehistoric Man in Western Europe,* Thames and Hudson, London.
    1976. *The Evolution of Palaeolithic Art,* in Fagan, B., ed., *Avenues to Antiquity, Readings from ScAm,* W. H. Freeman and Co., p. 66.

Lorblanchet, M., 1973. *La Grotte de Sainte-Eulalie,* in *GalliaPréh* t. 16, p. 3.

Luquet, G. H., 1930. *The Art and Religion of Fossil Man,* New Haven and London.

Pales, L., and St. Péreuse, M. de, 1969. *Les Gravures de la Marche, I. Félins et Ours,* Instit. de Préh. de l'Univ. de Bordeaux Mem. 7, see pp. 115–118.

Reinach, S., 1903. *L'Art et La Magie,* in *L'Anth* t. 14, p. 257.

Souriau, E., 1971. *Art Préhistorique et Esthétique du Mouvement,* in *Mélanges Offerts à A. Varagnac,* Ec. Pratiques des H. Etudes, Paris.

Spencer, B., and Gillen, F. J., 1899. *The Native Tribes of Central Australia,* London.

Stevens, A., 1975. *Animals in Palaeolithic Cave Art . . . ,* in *Ant* March, p. 54.

      Association of Animals in Palaeolithic Cave Art, in *Science and Archaeology* No. 16, p. 3.

Ucko, P. J., and Rosenfeld, A., 1967. *Palaeolithic Cave Art,* World University Library, London.

Quotations

1. Lartet, E., and Christy, H., 1864. *Sur des Figures d'Animaux Gravées ou Sculptées,* in *Revue Archéologique* t. 9.
2. Grosse, E., 1897. *The Beginnings of Art,* New York.
3. Boyd-Dawkins, W. G., 1880. *Early Man in Britain,* London, p. 214.
4. Cartailhac, E., 1878. *Matériaux pour l'Etude de l'Histoire Naturelle et Primitive de l'Homme,* Paris, p. 4.
5. Cartailhac, E., 1889. *La France Préhistorique, Bibliog. Scient. Int. 48,* Paris, pp. 79–81.
6. Souriau, E., 1971, p. 704.
7. Reinach, S., 1903, p. 257.
8. Ibid.
9. Laming-Emperaire, A., 1971, pp. 549–550.
10. Leason, P. A., 1939, p. 53.
11. Ibid., p. 59.
12. Leroi-Gourhan, A., 1972. Discussion in *Santander Symposium, Actas del Symp. Int. de Arte Preh.,* Santander, p. 304.
13. Ucko, P. J., and Rosenfeld, A., 1967, p. 239.
14. Leroi-Gourhan, A., 1968, p. 34.
15. Ibid., p. 48.
16. Lorblanchet, M., 1974. *L'Art Préhistorique en Quercy,* Morlaas, St. Jammes, p. 100.
17. Laming-Emperaire, A., as quoted in Drouot, E., 1973. *Les Motivations de l'Art Paléolithique,* in *Etudes Préhistoriques,* Soc. Préh. de l'Ardèche, No. 4, p. 24.

Chapter Eleven

Human representations in Paleolithic art

For a masterly survey of the problem, see the 1976 book by Pales.

Abramova, Z., 1967. *Palaeolithic Art in the USSR,* in *ArcAnth* Vol. IV No. 2, p. 1.

Barandarian, I., 1971. *Hueso con Grabados Paleolíticos en Torre,* in *Munibe* Año XXIII, p. 37.

Bosinski, G., 1970. *Magdalenian Anthropomorphic Figurines at Gönnersdorf (Western Germany),* in *Bollettino del Centro Camuno di St. Pre.,* Vol. V, p. 57.

    and Fischer, G., 1974. *Die Menschendarstellungen von Gönnersdorf der Ausgrabung von 1968,* Koblenz.

Chollot, M., 1964. *Collection Piette,* M. des Ant. Nat., Paris, see p. 468 on "La Femme au Renne."

Clottes, J., and Cérou, E., 1970. *La Statuette Féminine de Monpazier,* in *BSPF* t. 67 Et. et Trav., p. 435.

Pales, L., 1976. *Les Gravures de La Marche, II—Les Humains,* Ophrys, Paris.

    and St. Péreuse, M. de, 1964. *Une Scene Gravée Magdalénienne,* in *Objets et Mondes* t. 4, p. 77.

    and St. Péreuse, M. de, 1968. *Humains Superposés de La Marche,* in *La Préhistoire, Problèmes et Tendances,* CNRS, Paris, p. 327.

Péricard, L., and Lwoff, S., 1940. *La Marche . . . ,* in *BSPF* t. 37, p. 155.

Ucko, P. J., 1968. *Anthropomorphic Figurines,* Royal Anth. Inst. Occ. Paper No. 24, London, see p. 409.

    and Rosenfeld, A., 1972. *Anthropomorphic Representations in Palaeolithic Art,* in *Santander Symposium,* p. 149.

Quotations

1. Pales, L., 1976, p. 93.
2. Graziosi, P., 1960. *Palaeolithic Art,* Faber, London, p. 46.
3. Ucko, P. J., 1968, p. 409.
4. Abramova, Z., 1967, p. 89.
5. As quoted in Pales, L., and St. Péreuse, M. de, 1969. *Les Gravures de La Marche, I. Félins et Ours,* Inst. de Préh. de l'Univ. de Bordeaux, Mem. 7, p. 14.
6. Ibid., p. 12.
7. As quoted in Pales, L., 1976, at the beginning of Observation 60.
8. Ibid., p. 111.

Chapter Twelve

Mobile art and the theories of Alexander Marshack

Perhaps the most interesting general review of an early stage of Marshack's research is to be found in the 1972 *CA* article and discussion.

Capitan, L., and Bouyssonie, J., 1924. *Limeuil . . .*, Inst. Int. d'Anth., Paris.

Clot, A., 1973. *L'Art Graphique Préhistorique des Hautes-Pyrénées,* Morlaas, St. Jammes, see p. 42 for Labastide.

King, A. R., 1973. Review of *Roots of Civilization* in *AmAnth* Vol. 75, p. 1897.

Leroi-Gourhan, Arl., 1973. *Le Paysage au Temps des Graveurs de la Grotte de La Marche,* in *Est. Dedicados al Dr Luís Pericot,* Univ. de Barcelona Inst. d'Arq. y Preh.

Lorblanchet, M., 1973. *La Grotte de Sainte-Eulalie,* in *GalliaPréh* t. 16, see pp. 285–291.

Marshack, A., 1969. *Polesini . . .*, in *Riv* XXIV (2), p. 219.

    1970. *Notations dans les Gravures du Paléolithique Supérieur . . .*, Inst. de Préh. de l'Univ. de Bordeaux, Mem. 8.

    1970. *Le Bâton de Commandement de Montgaudier . . .*, in *L'Anth* t. 74, p. 321.

    1972. *The Roots of Civilization,* McGraw, New York and Weidenfeld, London.

    1972. *Cognitive Aspects of Upper Paleolithic Engraving,* in *CA* Vol. 13, p. 445, and see Vol. 15, p. 328.

    1974. *Marshack's Critique of* The Roots of Civilization *and of its Review by A.R. King,* in *AmAnth* 76, p. 845.

    1975. *Exploring the Mind of Ice Age Man,* in *National Geographic* 147 (1), p. 62.

    1977. *The Meander as a System . . .*, in Ucko, P. J., ed., *Form in Indigenous Art,* Australian Inst. of Abo. St., London and New Jersey, p. 26.

Nilsson, M. P., 1920. *Primitive Time Reckoning,* Lund.

Pales, L., and St. Péreuse, M. de, 1969. *Les Gravures de La Marche, I. Félins et Ours,* Inst. de Préh. de l'Univ. de Bordeaux, Mem. 7, see pp. 111–118 on "art school" theory.

Pericot García, L., 1942. *La Cueva del Parpallo,* CSIC, Madrid.

Quotations

1. Pales, L., and Tassin, M., 1968. *Humains Superposés de La Marche,* in *La Préhistoire, Problèmes et Tendances,* CNRS, Paris, p. 327.
2. Brown, G. B., 1928. *The Art of the Cave Dweller,* London, p. 162.

3. van Gennep, E., 1925. *A Propos du Totemisme Préhistorique*, in *Actes du Cong. Int. d'Hist. des Réligions 1923*, Vol. 1.
4. Bosinski, G., 1970, *Magdalenian Anthropomorphic Figurines at Gönnersdorf (Western Germany)*, in *Bollettino del Centro Camuno di St. Pre.*, Vol. V p. 83.
5. Marshack, A., 1972, p. 16.
6. Marshack, A., 1974, p. 845.
7. Littauer, M. A., in *CA* Vol. 15 No 3 Sep. 1974, p. 328.

Chapter Thirteen

The post-glacial period and the invention of farming

There are many good accounts of the transition to farming economies, but see particularly Bender 1975. For adaptations during the Mesolithic period, the works by Clarke and Mellars, and the papers in Kozlowski 1973, are of special interest.

Bender, B., 1975. *Farming in Prehistory*, John Baker, London.
Binford, L. R., 1972. *Post-Pleistocene Adaptations*, in *An Archaeological Perspective*, Academic Press, New York and London, p. 421.
Chollot, M., 1964. *Collection Piette*, Mus. des Ant. Nat., Paris, see pp. 223 and 325 for Mas d'Azil finds.
Clark, G. A., 1974–5. *Excavations in the late Pleistocene Cave Site of Balmori . . .* , in *Quat* Vol. XVIII, p. 383 (on *concheros*).
Clark, J. G. D., 1972. *Star Carr: A Case Study in Bioarchaeology*, Addison-Wesley Modular Publications, Boston.
          1975. *The Earlier Stone Age Settlement of Scandinavia*, Cambridge University Press, Cambridge.
Clarke, D. L., 1976. *Mesolithic Europe: The Economic Basis*, in Sieveking, G., Longworth, I. H. and Wilson, K. E., eds., *Problems in Economic and Social Archaeology*, Duckworth, London, p. 449.
Evans, J. G., 1975. *The Environment of Early Man in the British Isles*, Elek, London.
Graziosi, P., 1973. *L'Arte Preistorica in Italia*, Sansoni, Florence.
Higgs, E. S., ed., 1972. *Papers in Economic Prehistory*, Cambridge University Press, Cambridge.
          1975. *Palaeoeconomy*, Cambridge University Press, Cambridge.
Kozlowski, J. K., ed., 1973. *The Mesolithic in Europe*, Warsaw.
Marshack, A., 1972. *The Roots of Civilization*, Weidenfeld, London and McGraw, New York, see pp. 341–364.
          1973. *Analyse Préliminaire d'une Gravure à Système de Notation de la Grotte du Tai*, in *EtPréh*, Soc. Préh. de l'Ardèche, No. 4, p. 13.

Megaw, J. V. S., ed., 1977. *Hunters, Gatherers and First Farmers Beyond Europe,* Leicester University Press. See important paper by A. J. Legge, p. 51.

Mellars, P. A., 1976. *Fire, Ecology, Animal Populations and Man,* in *PPS* Vol. 42, p. 15.

Mezzena, F., 1976. *Nuova Interpretazione delle Incisione Parietali Paleolitiche della Grotta Addaura . . . ,* in *Riv* XXXI–1, p. 61.

Piggott, S., Daniel, G., and McBurney, C. B. M., 1974. *France Before the Romans,* Thames and Hudson, London, see chapter by M. E. de Fonton, p. 61.

Struever, S., ed., 1971. *Prehistoric Agriculture,* Natural History, New York.

Suttles, W., 1968. *Coping with Abundance . . . ,* in Lee, R. B., and DeVore, I., *Man the Hunter,* Aldine, Chicago, p. 56. (Introduction to potlatch problem).

Tassé, G., 1970. *Les Petroglyphes du Bassin Parisien,* in *Valcamonica Symposium,* Actes du Symp. Int. d'Art Preh., Capo di Ponte, p. 95.

Ucko, P. J., and Dimbleby, G. W., eds., 1969. *The Domestication and Exploitation of Plants and Animals,* Duckworth, London.

Walker, M., 1971. *Spanish Levantine Rock Art,* in *Man* Vol. 197, p. 553.

Whyte, R. O., 1977. *The Botanical Neolithic Revolution,* in *HumEco* Vol. 5 No. 3, p. 209.

Quotations

1. Nougier, L. R., 1976. *L'Evolution Esthétique de l'Art Mobilier de Magdalénien Final des Pyrénées, Trav. de l'Inst. d'Art Préh. No. 15,* Toulouse, p. 295.
2. Clark, G., 1967. *The Stone Age Hunters,* Thames and Hudson, London, p. 92.
3. Evans, J. G., 1975, p. 90.
4. Bray, W., in Megaw, J. V. S., ed., 1977, p. 238.
5. Childe, V. G., 1951. *Man Makes Himself,* Mentor Books, New York, p. 23. (1st edtn., London, 1936).

Glossary

There are few references to archaeological methods in this book, mainly because of a wish to keep the text free of technicalities. For the reader interested in how some of the detailed conclusions and interpretations of archaeologists are reached, the following works may prove helpful.

Aitken, M. J., 1974. *Physics and Archaeology,* University Press, Oxford. (2nd edtn.).

Barker, P., 1977. *The Techniques of Archaeological Excavation,* London.

Brothwell, D., and Higgs, E. S., eds., 1969. *Science in Archaeology,* Thames and Hudson, London. (2nd edtn.).

Evans, J. G., 1978. *An Introduction to Environmental Archaeology,* Elek, London.

Hester, J. J., 1976. *Introduction to Archaeology,* Holt, Rinehart & Winston, New York.

Hole, F., and Heizer, R. F., 1973. *An Introduction to Prehistoric Archaeology,* New York. (3rd edtn.).

Renfrew, J. A., 1973. *Palaeoethnobotany,* Columbia University Press, New York.

Semenov, S., 1970. *Prehistoric Technology,* Adams and Dart, Bath.

For further references to the hunting peoples mentioned in this book, see Lee, R. B., and DeVore, I., eds., 1968. *Man the Hunter,* Aldine, Chicago.

Archaeomagnetic dating A method for determining the age at which certain materials acquired magnetic properties. The orientation and strength of the magnetic field of the sample are compared with the known changes in the earth's magnetic field over time. The sample must contain ferromagnetic material and must preserve the record of its ancient magnetism. The method is useful for establishing the firing date of some ceramics, or the age of fine deposits laid down in wet conditions, but is comparatively unreliable. See Tarling, D. H., 1975–6, in *WArch* Vol. 7 No. 2, p. 185.

Aurignacian One of the earliest cultures of the Upper Paleolithic period, named after the Aurignac cave in the French Pyrenees, dug by Lartet in 1860. Dates for French Aurignacian industries range from about 32,000 to 26,000 B.C., but in parts of central Europe, Asia and the Levant, earlier dates than this are evident.

Azilian Traditionally considered as the first post-glacial culture of southern France and northern Spain, identified by Piette in the course of excavation at Le Mas d'Azil in the central Pyrenees. In fact, the flint and bone tools are closely related to those of the Magdalenian culture. Dates in France are around 9,000 to 8,000 B.C.

Bâtons de commandement   The name applied to the pierced staves of antler or bone, widely distributed throughout the European Upper Paleolithic cultures. Their possible functions are discussed in Chapter 6.

Blades   Flint flakes of an elongated form that were often retouched to form a variety of tools. The common use of blades for toolmaking is a characteristic feature of the Upper Paleolithic cultures, although their manufacture was not unknown earlier.

Bolas   A hunting weapon consisting of round missiles attached to cords or thongs that are whirled in the air and aimed at the legs of fleeing animals in order to entangle them. Different types of bolas were employed by South American cowboys and by Eskimo hunters. It would appear to have been one of the earliest weapons devised. Stone spheres that may be parts of bolas are common finds in the later levels at Olduvai Gorge and in subsequent periods in Europe.

Bones—Animal and Human   Animal bones provide insights into the diet of prehistoric people, although there are important problems mentioned briefly in Chapter 5. For further information, see Uerpmann, H. P., 1972, in *WArch* Vol. 4, p. 307. For the study of human bones, see the indispensable book by Brothwell, D. R., 1972. *Digging Up Bones,* British Museum, 2nd edtn.

Bronze Age   A term for the period in which bronze metal-working first became common, although this ranged from about 3,000 B.C. in the Near East to about 2,000 B.C. in Britain. The picture of social change in Europe at this time is now so complex that the term is of doubtful value. See Renfrew, C., 1973. *Before Civilization,* Jonathan Cape, London, and Knopf, New York.

Calcite   Crystallized deposits of calcium carbonate.

Carbonic gas or acid   A gaseous compound of carbon dioxide and water (CH_2O_3) that attacks limestone—the carbonates are the salts of this compound.

Caribou   North American native name for the reindeer, considered to be the same species *(Rangifer tarandus)* as the Eurasian reindeer. More significant than regional distinctions appear to be the differences in habits between "woodland" and "tundra" forms of the same species. See Burch, E. S., 1972, in *AmAntiq* Vol. 37, p. 339.

Châtelperronian   Conventionally thought of as the first Upper Paleolithic culture in France, characterized by a blade tool retouched along one curving edge (the Châtelperron knife). Dated in France to about 32,000 to 31,000 B.C.

Concheros   A Spanish name for accumulations of marine shells discarded by prehistoric people (or "shell middens").

Concretion   A term used here to refer to the deposition of limestone in cave formations such as stalagmites or stalactites.

Core   A block of raw flint.

Cro-Magnon   A rock shelter site near Les Eyzies in the Vézère valley, where the celebrated discovery of five skeletons was made by Louis Lartet in 1868. These were among the first Ice Age remains of physically modern build to be found, and some authors persist in using the term Cro-Magnon to refer loosely to people of the Upper Paleolithic period.

Cup-marks   Simple depressions carved into stone surfaces—a basic element in prehistoric rock art traditions since their earliest beginnings.

Dingo   The native dog of Australia.

Domestication   The taming of animals or the cultivation of plants.

Erratic   A geological material that has been transported from its original source by natural agencies.

Gravettian   A culture of the Upper Paleolithic period named after the site of La Gravette in southwest France. It is used as an alternative term for the Upper Perigordian cultures of France, dating from roughly 26,000 to 20,000 B.C. In a wider and looser sense, it is also applied to many other flint-working traditions elsewhere in Europe of basically similar form.

Günz   The first major glacial episode of the Pleistocene, as defined in the classic Alpine system. Dating is problematic—one million years?

Hand-axe   Among the earliest standardized forms of stone implement and a common item in west European tool kits of the Lower and Middle Paleolithic. Fashioned from a core of flint, and probably used as an all-purpose cutting and butchering tool. There is no evidence for the hafting of hand-axes.

Hominid   A term applied to members of the zoological family *Hominidae.*

Homo sapiens sapiens   The zoological term for the species to which we belong. The now-common use of the term *Homo sapiens neanderthalensis* implies the acceptance of Neanderthal man as a member of our own species.

Hunting band   The basic unit of social organization among hunter-gatherers, conventionally thought of as a group of about twenty-five persons (perhaps half-a-dozen families) who trace their descent patrilineally or through the male line. But see Chapter 8 for different theories and observations.

Hunting blind   A screen of some kind devised by hunters to ambush game.

Ice Age   A long-term period of cold climatic conditions. The term Ice Age with capital letters in this book is used to refer only to the last Würm glacial period, dating from about 70,000 to 10,000 years ago.

Interglacial   An interval of mild climate between ice ages. During the Pleistocene there was a minimum of three such periods and an estimate of six may be nearer the mark. The exact definition of such episodes remains extremely difficult.

Interstadial   An interval of mild climate within an ice age or glacial period. At least seven such phases are recognized by some researchers in France during the period of the Upper Paleolithic cultures alone.

Iron Pyrites   A sulphide of iron.

Magdalenian   The last and greatest of the classic French sequence of Upper Paleolithic cultures, with dates from about 16,000 to 9,000 B.C. (although antecedents before the Solutrean period are possible). It is named after the site of La Madeleine on the Vézère, first excavated in 1863. There were extensions of the Magdalenian culture in Spain, Belgium, Czechoslovakia, Britain, and notably in north and west Germany and Switzerland towards the end of the period.

Maglemose A name given to the early post-glacial cultures of northern Europe, embracing much of Britain, the southern Baltic, northern Germany and western Russia. These diverse regions are linked by the common character of items of equipment such as harpoon heads of bone or antler, although the usefulness of such a "blanket" term is debatable.

Manganese dioxide A naturally occurring black mineral, which seems to have been the black pigment most often employed by the cave artists.

Marques de chasse A French name for the series of abstract notches and marks that are commonly found on bone and antler objects from the Upper Paleolithic period. "Notations" is the word used by Alexander Marshack to describe them. See Chapter 12.

Megalithic tomb A funerary structure made of large stone slabs of the type erected by Neolithic peoples throughout western Europe from about 4,000 B.C. onwards.

Mindel The second and most severe of the Pleistocene glaciations. There are enormous differences in the estimates of the dating of this and other ice ages. It may have been as late as 320,000 to 220,000 years ago.

Mobile art Decorated objects that could have been picked up and transported from one site to another.

Moraine An accumulation of debris carried down by a glacier.

Mousterian The cultures of the Middle Paleolithic, named after the site of Le Moustier on the Vézère, first explored by Lartet and Christy in the 1860s. The term embraces a wide number of different industries on flakes, ranging throughout Europe, north Africa and the Near East. Within France four basic variants are recognized—the Mousterian of Acheulian tradition, the Charentian, the Typical and the Denticulate, dating at least from 70,000 B.C. to well after 40,000 B.C. See Chapter 3.

Neanderthal A general term often used imprecisely to refer to people of the Middle Paleolithic, before the advent of modern man. The word "Classic" is often added to denote the physically distinctive group of prehistoric west Europeans dating to the earlier part of the last glaciation. See Chapter 3.

Neolithic   The New Stone Age, marked by the coming of agriculture, dating in the Near East to at least 6,000 B.C. and to somewhat later periods in Europe and the Americas.

Noaillan   A name given to a distinctive variation of the normal toolmaking practices of the Perigordian cultures in France and dating to about 23,000 B.C.

Notations   Marks or symbols that record numbers or quantities, such as time intervals. Used by Alexander Marshack to refer to the puzzling marks on bone and antler objects, certain of which are claimed to be a form of calendrical marking. See Chapter 12.

Ocher   A naturally occurring mixture of clay and hydrated iron oxide, also known as haematite, that ranges from shades of light yellow to red and brown. Variations in the color may be produced by heating the ocher. This was apparently the pigment most commonly used by the cave artists.

Paleolithic—Lower, Middle and Upper   The Old Stone Age. Originally prehistory was divided simply into the Neolithic, with agriculture, and the Paleolithic, without. The Paleolithic is now subdivided into the Lower, with pebble tool and hand-axe industries; the Middle, with the Mousterian; and the Upper Paleolithic, which includes the industries of the cave artists up to the end of the Ice Age. The nature and timing of the divisions between these periods are far from clear.

Pebble tools   The most primitive type of stone tool known, consisting merely of a sharp edge created on a pebble by flaking from one or two sides. At present the earliest such tools are recognized at sites in East Africa at about two and one-half million years.

Perigordian   One of the cultures of the French Upper Paleolithic, named after the Périgord region of the southwest. Its many different stages date roughly from about 26,000 to 20,000 B.C. See also the term Gravettian.

Permafrost   Conditions of permanently frozen subsoil.

Plaquette sanctuary A term used in this book to refer to occupation sites with mobile art in the form of an unusual number of engraved stone blocks.

Pleistocene A geological time division or epoch, covering much of the last two to three million years.

Pollen The tough skin or exine of pollen ensures its preservation in certain conditions, best of all in anaerobic peats or lake deposits. By magnifying samples, the characteristic shapes of different pollens can be identified by eye. The quantity, proportions and range of species in a sample are a guide to the vegetation present. But there are problems, such as pollen blown in from non-local sources, differential production by various types of tree, factors of preservation, climate, topography, etc. See Dimbleby, G. W., 1978. *Plants and Archaeology,* John Baker, London. (2nd edtn.).

Potlatch Ceremonial gathering and gift-giving of the northwest coast Indians of America. For an introduction to the potlatch, see Suttles, W., 1968, in Lee, R. B., and De Vore, I. eds., 1968. *Man the Hunter,* Aldine, Chicago, p. 56.

Pyrenean izard The Pyrenean chamois.

Radiocarbon dating One of the most useful dating techniques known to archaeologists. All organic material, such as charcoal, wood, bone or antler, contains small amounts of the radioactive substance known as carbon 14, which decays gradually at a known rate. The measurement of the amount of carbon 14 present in a sample allows a calculation of its age to be made, within certain margins of error. Unfortunately, the method becomes increasingly unreliable up to the period of the beginning of the European Upper Paleolithic, around 40,000 B.C. As this book was going to press, new methods of laser enrichment were announced that may extend the usefulness of radiocarbon dating back as far as 100,000 years. This promises an enormous gain in knowledge of the chronology of the early Ice Age. See *Nature,* November 16th 1978, 276, 255, and Renfrew, C., 1973. *Before Civilization,* Jonathan Cape, London, and Knopf, New York.

Riss The third, penultimate European glaciation, as it is known in the classic Alpine system. The dating may range from about 200,000 to 100,000 B.C. at a very rough estimate.

Rock shelter   An overhang in a cliff face affording shelter for a group of prehistoric hunters.

Shaman   Priest or priest-doctor, formerly widespread among the tribes of northern Asia, whose powers were thought to exert an influence on the spirit world.

Soils   The analysis of the character of soil samples can provide general indications of climatic conditions. Sequences of soil changes compared between sites in a particular region can provide an approximate means of dating them. See Evans, J. G., 1978. *An Introduction to Environmental Archaeology,* Elek, London.

Solutrean   The briefest and most restricted of the Upper Paleolithic cultures in the classic French sequence, named after the great open-air settlement of Solutré in the Saône valley of southeast France, first dug in 1865. Dates in France range from about 20,000 to 16,000 B.C. at the maximum. In its later stages the Solutrean is known in parts of Spain and Belgium. See Smith, P., 1976, in Fagan, B., ed., *Avenues to Antiquity, Readings from ScAm,* W. H. Freeman and Co., San Francisco.

Spear-thrower   A shaft with a notch or hook on one end that engages with the butt of a spear. The spear is laid flat against the spear-thrower, which is then jerked sharply forward by the hunter. This device, common to many hunting peoples throughout the world, increases the leverage and range available to the thrower by effectively adding to the length of the arm.

Stalactite   A deposit of limestone from the ceiling of a cave, usually in the shape of an icicle.

Stalagmite   A deposit like a stalactite, but usually on the floor of a cave.

Tell   The Arabic word for an artificial mound, usually covering the remains of a settlement.

Tundra   The flat, treeless regions of northern Eurasia and Canada, characterized by Arctic climate, vegetation and fauna.

Venus   A nickname widely and perhaps inappropriately used for female statuettes of the Upper Paleolithic period. See Chapter 11.

Wear-marks Microscopic study of the patterns of wear on the surfaces of flint tools provides useful indications of their possible function. The damage originally inflicted on the flint will depend on the duration, the force applied, the speed, working position, tool shape and the material involved. Experimental work is usually undertaken to reproduce comparable patterns of wear for comparison. For some of the problems, see Keeley, L. H., 1974, in *WArch* Vol. 5.

Würm The "Ice Age" of the title. The fourth and last major glacial period in the classic Alpine sequence, with dates ranging from about 75,000 to 9,000 B.C. at most.

Photo Credits

Altuna, Dr. J., courtesy of, 167 (bottom), 192, 208, 210. These photographs originally appeared in *Munibe*. Sociedad Aranzadi, San Sebastian.

American Museum of Natural History, courtesy of, 7, 151, 222 (left and center).

Anthro-Photo, Cambridge, Mass., by I. DeVore, 129, 132, 153.

Arambourou, R., courtesy of, 82, 83.

Australian Information Service, London, 5, 16, 112, 170, 171, 201.

Barrière, C., 247 (reproduced from *L'Art Pariétal de la Grotte de Gargas*. British Archaeological Reports, Supp. Series No. 14 1975, fig. 84.

Bogoras, W., 123 (reproduced from *The Chuckchee*. Mem. Am. Mus. Nat. Hist., Vol. VII 1907, pp. 85 and 163).

Breuil, H., 9 (Pl. LX, *Les Cavernes de la Région Cantabrique*. 1906), 80 (from *Beyond the Bounds of History*. P.R. Gawthorn 1949, p. 80).

British Museum, reproduced by courtesy of the Trustees, 107 (top), 226.

Brooke, A. de C., 120 (from *A Winter in Lapland and Sweden*. 1827, p. 520).

Brown, G.B., 73, (from *The Art of the Cave Dweller*. 1928, p. 31).

Bulletin de la Société Préhistorique Française. 180 (bottom), 194.

Caisse Nationale des Monuments Historiques, 188, © ARCH. PHOT. PARIS/ SPADEM.

Canadian Government Photo, 115, National Film Board of Canada, 104 (top) and 138. By courtesy of the Canadian Information Service, London.

Clottes, J., by courtesy of, 180 (bottom, plan by J. Clottes and R. Bégouën), 194, 196 (photographs by J. Clottes and R. Simmonet).

Coleman, Bruce, Ltd., 24, by Bob Campbell.

Combier, J., by courtesy of, 104 (bottom, by G. Taupenas), 292.

Congr. Int. d'Anthrop. et d'Arch. Préhist.. 178 (from t. l, 1912, opp. p. 496).

Dominguez, E., 101 (top), 205.

Figuier, L., 128, 198 (from *L'Homme Primitif*. 1870, figs. 39 and 68).

Fine, D., 285.

Fournier, Alain, 167 (top), 176, 177, 213.

Gallia Préhistoire. 196, 256, 257, 286.

Hadingham, Evan, frontispiece, 39 (right), 79 (courtesy of F. Champagne), 101 (bottom), 144, 145, 146, 172, 190–191, 200, 212, 227, 253, 260, 261.

Hooper, Alex, and David Collison, 288, and Evan Hadingham, 26, 31.

Levy, Tom, 46, 47.

Lorblanchet, M., 256, 257 (reproduced from *Gallia Préhistoire*. t. 16, 1973, pp. 26 and 27).

Lumholtz, C., 136, 274 (from *Among Cannibals*. 1889).

Lumley, H. de, by courtesy of, and the Lab. de Pal. Hum. et de Préh., Université de Provence, 29, 30, 51, 88.

Musée des Antiquités Nationales, St. Germain, by courtesy of, 110, 161.

Musée de l'Homme, Paris, 38, 39 (left, by J. Oster), 44, 57 (by H. Dartin), 62, 63 (by P. Poulain), 67, 111 (by J. Oster), 118 (by Tollmanoff), 141 (by J. Oster), 185 (by G. Richard), 206 (by J. Oster), 223, 235, 237 (by J. Oster), 245.

Nationalmuseet I, København, 269, by Lennart Larsen.

Novosti Press Agency, London, 13, 15, 71, 76, 77, 107 (bottom), 113, 121.

Osborn, H.F., 35 (from *Men of the Old Stone Age*. 1915).

Pales, L., by courtesy of, 206 (reproduced from *Les Gravures de La Marche*. t. l, Instit. de Préh., Université de Bordeaux, 1969, obs. no. 7, Pl. 19a and b), and the following illustrations (all from *Les Gravures de la Marche*. II—*Les Humains*. 1976) by permission of Dr. Pales and Editions Ophrys: 231 (fig. 4, no. 1–13), 234 (156), 235 (159), 236 (115a and b), 238 (38a).

Peabody Museum, Harvard University, 103.

Piette, E., 9 (from *La Revue Préhistorique*), 265 (from *L'Art Pendant l'Age du Renne*, 1907).
Roussot, Alain, 248.
Schmidt, R.R., 222 (right, from *The Dawn of the Human Mind.* 1936, p. 123).
Soprintendenza alle Antichita, Palermo, 262.
Vertut, J., 180 (top), © Collection BEGOUËN, Cliché J. Vertut.
Wheeler, B., University of Sheffield, 17, 86, 87.
Wildgoose, M., 147.

Text Figure Credits

10, after Ucko and Rosenfeld 1967, fig. 5 and Lorblanchet 1973, fig. 1.
15 (right), after Pidoplichko and McBurney 1976.
19, after Manley 1953 and Goudie 1977 p. 4.
20 (top), after Segota 1966 and Goudie 1977 p. 33; (bottom), after Frenzel 1973 and Goudie 1977 p. 171.
21, after Flint 1971, Kaiser 1969 and Goudie 1977 p. 40.
33, after P. Laurent in Bordes 1969 and A. Marshack 1977.
40, after M. Boule and H.V. Vallois 1957.
43, after Chernysh 1969 and McBurney 1976.
59 (top), after E. Bonifay 1965; (left) after A.C. Blanc 1956; (right) after S. Giedion 1957 Vol. I.
63, after A. Leroi-Gourhan 1961.
65, after Jelinek 1975.
84, after G. de Mortillet 1874.
90, after chart in *BSPF* t. 71 no. 5 Mai 1974, p. 135.
98–99, after M. Gábori 1976, fig. 59.
103, after J. Clottes in H. de Lumley, ed., 1976, pp. 1215 and 1219.
108, after Ormeaux 1889.
110, all after M. Chollot 1964 except (bottom right) after J. Omnès 1977.
143 (left), after Musée de l'Homme, and (right) after Vallois 1971 in *BSPF* t. 68.
147, after C. Barrière 1975, fig. 17.
158 (left), after Boriskovskij 1958, fig., 3, and (right) after L. Banesz 1976 in *UISSP IX Congr.*, fig. 5.
161 (bottom), after Passemard 1944 and P. Graziosi 1960.
165, after Sieveking 1976 and Clottes 1976 in H. de Lumley, ed., p. 1219.
168, after Sieveking 1976.
175, after Hooper and Collison 1976, in *Gallia Préhistoire.* p. 222.
183, after Breuil and Bégouën 1958, Pl. XX.
199, after Breuil and St. Périer 1926.
215, after A. Leroi-Gourhan 1968.
221, after Clottes and Cérou in *BSPF* t. 69 1970 p. 441.
228, after I. Barardarian 1975, in *La Prehistoria en la Cornisa Cantabrica.*
241 (left), after Bouyssonie 1920, and (right) after Cartailhac in de Mortillet 1885 p. 64.
243, after Bosinski 1970, figs. 52, 53.
252, after Breuil and St. Périer 1926.
254, after Radmilli 1973.
261, after Mezzena 1977.
268–269, after Mazonowicz 1970.

272–273, after Arl. Leroi-Gourhan and M. Girard in H. Laville and J. Renault
   Miskovsky 1977.
276, after M. O'Malley in *Current Archaeology.* 1978, p. 118.
289, after Breuil 1911 and Clot 1973.
295, after M.G. Levin and L.P. Potapov 1969.

Maps of Decorated Caves

DORDOGNE, after M. Sarradet 1975 Vol. 6, Ed. du Centro Camuno, Vol. 1.
QUERCY, after M. Lorblanchet in *BSPF* t. 68 1971 p. 309, fig. 15.
PYRENEES, after Sieveking 1976 fig. 1, and Clottes 1976 p. 1219, fig. 4.
N. SPAIN, after various maps in *La Prehistoria en La Cornisa Cantabrica.* 1975.
UKRAINE, after D. Nat 1971, fig. 2.

Index

Illustrations are indicated by **bold** page numbers.

Fragment of the sculptured relief at
Angles-sur-l'Anglin, Vienne, western France.

Date Due

BRODART, CO.   Cat. No. 23-233-003   Printed in U.S.A.